Environmental Science and Engineering
Subseries: Environmental Science

Series Editors: R. Allan • U. Förstner • W. Salomons

S. Nayak · S. Zlatanova (Eds.)

Remote Sensing and GIS Technologies for Monitoring and Prediction of Disasters

Springer

Editors
Dr. Shailesh Nayak
ISRO Government of India
Ahmedabad-380015
Jodhpur Tekra
India
director@incois.gov.in

Dr. Sisi Zlatanova
Delft University of Technology
OTB Research Inst. Housing
Urban & Mobility Studies
2600 GA Delft
Netherlands
s.zlatanova@tudelft.nl

ISBN: 978-3-540-79258-1 e-ISBN: 978-3-540-79259-8

Environmental Science and Engineering ISSN: 1863-5520

Library of Congress Control Number: 2008930076

Cover Design: Integra Software Services Pvt. Ltd.

Printed on acid-free paper

9 8 7 6 5 4 3 2 1

springer.com

Contents

Contributors

François Anton: Department of Informatics and Mathematical Modelling, Technical University of Denmark, Lyngby, Denmark, fa@imm.dtu.dk

Rajarathinam Arangarasan: The Raj Organization LLC, Las Vegas, Nevada, USA, raj@rajarathinam.com

Anjali Bahuguna: Space Applications Centre (ISRO), Ahmedabad, India, anjali@sac.isro.gov.in

Larry Biehl: Rosen Center for Advanced Computing, Purdue University; Purdue Terrestrial Observatory, West Lafayette, Indiana, USA, biehl@purdue.edu

Erik van Borkulo: Geodan, Amsterdam, The Nederlands, erik.van.borkulo@geodan.nl

Henrike Brecht: LSU Hurricane Center, Louisiana State University, LA, USA, henrike@hurricane.lsu.edu

Alok Chaturvedi: Purdue University; Purdue Homeland Security Institute, West Lafayette, Indiana, USA, alok@purdue.edu

Arta Dilo: OTB Research Institute for housing, urban and mobility studies, Delft University of Technology, Delft, The Netherlands, a.dilo@tudelft.nl

Souleymane Fall: Earth and Atmospheric Sciences, Purdue Terrestrial Observatory, West Lafayette, Indiana, USA, sfall@purdue.edu

Donald Fortin: Direction des opérations, Ministère de la Sécurité publique du Québec, St-Jean-sur-Richelieu, Québec, Canada, donald.fortin@msp.gouv.qc.ca

Steven Fruijtier: Geodan, Amsterdam, The Netherlands, steven@geodan.nl

Harmen Hofstra: Vrije Universiteit, Amsterdam, The Netherlands, h.hofstra@xs4all.nl

Yves Gauthier: Laboratoire de télédétection, Institut national de recherche scientifique - Eau, Terre et Environnement, Québec City, Québec, Canada, yves_gauthier@inrs-ete.uquebec.ca

Saibal Ghosh: Geological Survey of India, Salt Lake, Kolkata, India, ghosh@itc.nl

Swati Gehlot: Max-Planck Institute of Meteorology, Hamburg Germany, swati.gehlot@zmaw.de

Timir B. Ghoshal: Geological Survey of India, Salt Lake, Kolkata, India, tbghosal@gmail.com

Ramon F. Hanssen: Aerospace Engineering, University of Technology, Delft, The Netherlands, R.F.Hanssen@tudelft.nl

Viswanath S. Hegde: SDM College of Engineering and Technology, Dharwad, India, vshegde2001@yahoo.com

Jan Herrmann: Department for Geography, Ludwig-Maximilians-Universität München, Munich, Germany, jan.herrmann@lmu.de

Norman Kerle: Department of Earth Systems Analysis, International Institute for Geonformation Science and Earth Observation (ITC), Enschede, The Netherlands, kerle@itc.nl

Gengsheng Liang: Department of Geodesy and Geomatics Engineering, University of New Brunswick, Fredericton, Canada, c1g68@unb.ca

Krishna Madhavan: Clemson University, Clemson, South Carolina, USA, cm@clemson.edu

Robert M. Mikol: Geographic Information Network of Alaska, University of Alaska, AK, USA, rmikol@gi.alaska.edu

Darka Mioc: Department of Geodesy and Geomatics Engineering, University of New Brunswick, Fredericton, Canada, dmioc@unb.ca

Shailesh Nayak, Indian National Council for Ocean Information Services (INCOIS), Hyderabad, India, director@incois.gov.in

Dev Niyogi: Purdue University and Indiana State Climatologist, West Lafayette, Indiana, USA, climate@purdue.edu

Joseph Quansah:, Department of Agricultural & Biological Engineering Purdue University, jquansah@purdue.edu

Ajay S. Rajawat: Space Application Centre, Ahmedabad, India, asrajawat@hotmail.com

Narayanaswamy Rajendran: Geological Survey of India, Op. Karnataka & Goa, Bangalore, India, n_rajendran@yahoo.com

Gilbert L. Rochon: Rosen Center for Advanced Computing, Purdue University; Purdue Terrestrial Observatory, West Lafayette, Indiana, USA, rochon@purdue.edu

Henk Scholten: Vrije Universiteit, Amsterdam, The Nederlands, hscholten@feweb.vu.nl

Alessandra Scotta: Geodan, Amsterdam, The Netherlands, alessandra.scotta@geodan.nl

Charles Siegel : TGIS Technologies inc., Chelsea, Québec, Canada, charles.siegel@tgis.ca

David Stevens: United Nations Office for Outer Space Affairs, Vienna, Austria, david.stevens@unvienna.org

G. Shalini: Global Academy of Technology, Rajarajeshwari Nagar, Bangalore, India, shal-sham@rediffmail.com

Mopur Surendranath: Geological Survey of India, Bandlaguda, Hyderabad, India, msurendranath@gmail.com

Barandi Widartono: Cartography and Remote Sensing Department, Faculty of Geography, Gadjah Mada University, Yogyakarta, Indonesia, barandi_sw@yahoo.com

Sisi Zlatanova, OTB Research Institute for housing, urban and mobility studies, Delft University of Technology, Delft, The Netherlands, s.zlatanova@tudelft.nl

Introduction

Sisi Zlatanova and Shailesh Nayak

Natural and anthropogenesis disasters cause widespread loss of life and property and therefore it is critical to work on preventing hazards to become disasters. This can be achieved by improved monitoring of hazards through development of observation systems, integration of muti-source data and efficient dissemination of knowledge to concerned people. Geo-information technologies have proven to offer a variety of opportunities to aid management and recovery in the aftermath. Intelligent context-aware technologies can provide access to needed information, facilitate the interoperability of emergency services, and provide high-quality care to the public.

Disaster management poses significant challenges for real-time data collection, monitoring, processing, management, discovery, translation, integration, visualisation and communication of information. Challenges to geo-information technologies are rather extreme due to the heterogeneous information sources with numerous variations: scale/resolution, dimension (2D or 3D), type of representation (vector or raster), classification and attributes schemes, temporal aspects (timely delivery, history, predictions of the future), spatial reference system used, etc.

There is a need to continuously discuss the state of the observing systems and integration of effective monitoring of disasters, development of predictions systems, integration and analysis of geo-information. Recognising the importance of use of geo-information in disaster management, several universities (Delft University of Technology, VU University Amsterdam, The Netherlands; University of Waterloo, Canada), international organisations (ISPRS, UNOOSA, EU, ICA, FIG, OGC) and vendors (Bentley, Intergraph, Oracle, PCI) have taken the initiative to organise an annual symposium, which aims at uniting the efforts of researchers, developers, data providers and users from different countries and continents. The symposium was organised first in Delft, The Netherlands (March, 2005). Three more symposia were organised under the coordination of the ISPRS WGIV/8: Goa, India (September 2006), Toronto, Canada (2007) and Harbin, China (August, 2008).

The second symposium concentrated on natural disasters as the general theme was 'Remote Sensing and GIS Techniques for Monitoring and Prediction of Disasters'. It was organised by the Indian Society of Remote Sensing, ISPRS, ISRO, UNOOSA, FIG, EC, AGILE, ICA and Delft University of Technology on 25-26th of September 2006, Goa, India. The two-day symposium has accommodated 60 participants from 12 countries.

From the originally 96 submitted abstracts (from 28 countries), 46 full papers were received. The papers were presented in 6 oral sessions and one poster session in the first day. The symposium was closed with a panel session devoted to providing timely geo-information, quality of data, use of technical expertise after a disaster and involvement of geo-specialist in efforts to predict and mitigate disasters.

There are practically no doubts about current status of technology in providing spatial data to end users. Global navigation satellites and Earth observation satellites have largely demonstrated their flexibility in providing data for a broad range of applications: weather forecasting, vehicle tracking, disaster alerting, forest fire and flood monitoring, oil spills detection, desertification spread monitoring, crop and forestry damage assessment. Monitoring and management of recent natural disasters have also benefited from satellite imagery, such as the Indian Ocean tsunami in 2004, floods (Austria, Romania, Switzerland, and Germany in 2005), hurricanes (USA in 2005), forest fires (Portugal, France in 2005), earthquakes (Pakistan in 2005, Indonesia in 2006), etc.

However, it is recognised that effective utilisation of satellite positioning and remote sensing in disaster monitoring and management requires research and development in numerous areas: data collection, access and delivery, information extraction and analysis, management and their integration with other data sources (airborne and terrestrial imagery, GIS data, etc.) and data standardization. Establishment of Spatial Data Infrastructure at national and international level would greatly help in supplying these data when necessary. In this respect legal and organisation agreements could contribute greatly to the sharing and harmonisation of data.

Quality of data in case of disaster is still a tricky issue. Data with less quality but supplied in the first hour might be of higher importance in saving lives and reducing damages compared to trusted, high quality data but after two days. Apparently a balance should be found in searching and providing data as the general intention should be increased use of accurate, trusted data.

Charters and international organizations have already launched various initiatives on the extended utilization of satellite positioning and remote sensing technologies in disaster monitoring and management. For example, the International Charter is often given as a good example of availability of data and expertise after a disaster, but still the coordination between the different initiatives at local and international level is considered insufficient. This observation is especially strong for developing countries, although some authorities in developed countries (e.g. USA in the case of Hurricane Katrina) also fail to react appropriately. Capacity building needs to be further strengthened and the governments must be the major driving factor in this process. Related to this is the role of the geo-specialist in disaster management. Geo-specialist are not directly involved in emergency response, e.g. training together with first responders or preparing monitoring and mitigation programs, but there is high understating of closer work with users.

The Second Symposium has clearly revealed regional specifics in disaster management. While the symposium in Europe addressed Spatial Data Infratsructures and cooperation between different rescue units as major challenges, the symposium in India discussed mostly availability and processing of data and put emphasis on early warning systems, realizing that the national SDI for disaster management either do not exist or are at a very early stage.

The chapters of this book reflect some of the topics mentioned above. The efforts of many researchers over the past four years to continue research and development in the area of spatial data integration for effective emergency services and disaster management have also provided guidance and inspiration for the preparation of this book.

This book consists of 14 chapters organised in three parts. The readings in this book outline major bottlenecks, demonstrate use of remote sensing technology, and suggest approaches for sharing and access of information in various stages of disaster management process.

Part 1: Use of geo-information technology in large disasters.

The first chapter of Kerle and Widarontono elaborate on use of geo-information during the earthquake on 27 May 2006 in the Yogyakarta area, Indonesia. The authors provide numerous chronological details on the work of the different local and national organisations involved and the use of remote sensing data. This particular disasters is an excellent illustration of the works completed after the activation of the International Charter 'Space and Major Disasters'. Thanks to the almost immediate activation of the Charter, much satellite information could be quickly provided in the first two days. The authors also address some issues that need further improvement such as prices, availability of high resolution data, etc.

The second chapter is devoted to the lessons learned from the Katrina hurricane. The author Henrike Brecht has participated in the emergency response activities immediately after the water flooded the city of New Orleans. The personal observations of the author are organised in five groups of lessons namely management, technology and infrastructure, data, operational (and workflow) and map products. Clearly, many improvements have been observed in providing and use of geo-information comparing to any other disaster in USA, but problems still exist. The chapter provides a very good overview on bottlenecks and failures largely contributing to the 'what went wrong' issue. Interestingly, the lessons learned are very similar to the 9/11 experiences.

The third chapter addresses the damages on the fauna and flora on the Indian cost after the Tsunami, December 2004. Shailesh Nayak and Anjali Bahuguna present their elaborated study on the impact on major ecosystems applying high-resolution satellite imagery. As illustrated in the chapter, the applied methodology has helped to estimate the loss and help in the rehabilitation process. The damage to ecosystem (especially the coral reef and the mangroves) is critical as it directly affects fishery recourses of coastal communities.

After the first three chapters on use of remote sensing information for monitoring and damage assessment, Chapter 4 elaborates on a new initiative for developing countries that have to further ease access and sharing of satellite data. David Stevens elaborated on the tasks and activities of the United Nations Platform for Space-based Information for Disaster Management and Emergency Response (UN-SPIDER), established in December 2006 as a program of the United Nations Office for Outer Space Affairs. Through presenting recent major meetings, conferences and assemblies, and summarizing the most important activities of various organisations, the author motivates the work of the new program.

Part 2: Remote sensing technology for disaster monitoring

The second part of this book consists of five chapters all presenting remote-sensing technologies (satellite imagery, radar technology, Global Positions Systems) applied for various hazards or phases of disaster management process.

Chapter 5 is a collaborative work of eight universities and organisations and present a broad overview on need of different technology for monitoring of hazards, response to disaster, recovery and mitigation. The authors discuss availability of remote sensing data (illustrated with useful web links), provide practical examples from case studies and report software developments within the participating organisations. Special attention is given to dynamic integration of data for geo-visualisation in virtual environments and on hand-held. The chapter concludes with thoughts about a well-recognised need for an appropriate geo-education for disaster managers.

Hedge, Shalini, Nayak and Rajawat present a satellite-image based methodology for monitoring of pollution in the shore water. According to the authors, the pollution dispersion in the near shore water is highly complex and dependent on a number of factors. Ocean Colour Monitor patterns as well as other satellite data products have helped to successfully trace sediments dispersion and understand sediment dynamics through the different seasons.

In the following chapter, Surendranath, Saibal Ghosh, Ghosal and Rajendran address mapping of landslides in the Darjeeling Himalayas in East India. Again, this chapter is an excellent illustration of use of remote sensing data for monitoring of hazards. The authors are confident that some conventional methods based on aerial photogrammetry and manual inspection can successfully be replaced by high-resolution satellite images. The chapter present details on the methodology for the derivation of accurate DEM from topographic maps, IRS pan stereoscopic satellite imagery and freely available Shuttle Radar Topography Mission elevation data. Their method is especially suitable for highly rugged hilly areas, which are constantly under the highly dynamic and active erosion processes.

Chapter 8 discusses the potential of Persistent Scattered Interferometry for detection and monitoring of land subsidence. This technology reveals high cost effectiveness compare to conventional geodetic techniques. Besides the applicability of radar technology for monitoring of deformations, this research stresses the need of incorporating supplementary geo- information sources for an improved interpretation. Swati Gehlot and Ramon Hanssen report very promising results of applying this technology in the city areas in the Netherlands.

The last chapter in this Part 2 presents an extended procedure for GPS data collection. The improved procedure makes use of special waypoint protocol. Robert Mikol discusses the waypoint naming in detail (and the consequent organisation in a database) and illustrates its applicability for rapid data collection in case of oil spill. Though the DBMS has been never used during oil spill and subsequent cleanup, the idea was accepted as successful for data collection under limited financial resources.

Part 3: System architectures for access of geo-information

This part presents different approaches for management, access and sharing of geo-information for disaster management. Though not specifically concentrated on remote sensing data, the presented systems can easily be used if remote sensing imagery is available.

Jan Herrman addresses the very important issue of access and sharing of data through web services. Access control and security (protection of information) are especially important to enforce restricted access to protected spatial data or to declare views on the relevant data for certain users/roles. The author provides a overview on existing technology and elaborates on the advantages of Geo OASIS's eXtensible Access Control Markup Language (GeoXACML).

Chapter 11 elaborates further on a Spatial Data Infrastructure for disaster management in the Netherlands. The presented system architecture is a typical example of thin client-server architecture, which should be able to serve any type of user on the field or in the commando center. The implemented services are context-oriented and follow recent standardization developments toward chaining of generic services. A spatio-temporal model for management of operational data is one of the few attempts worldwide to manage emergency operational data in DBMS.

Charles Siegel, Donald Fortin and Yves Gauthier report on their system for cooperation and collaboration during emergencies. As discussed in the chapter, real-time contact and making available all the data to all the participants in an emergency is considered a key component in every command and control system. The developed system allows live Internet geo-collaboration, which is in use in civil security operating in the Québec Ministry of Public Security. The presented case studies come from real emergency management situations.

Chapter 13 presents yet another approach to access and visualize data over Internet, this time for flood management. The authors extensively discuss the decisions taken in development of the client-server architecture, data model for management of flood information, street/road networks and other spatial information. This approach convincingly illustrates the advantage of storing and managing information in DBMS: integrated spatial analysis can readily be performed at the server. A light web-application allows for visualization and inspection of performed analysis.

The last chapter presents a usability study about new type of hardware, i.e. Multi Tangible Tabletop User Interface, for its applicability in disaster management. The tangible table does not require use of mouse and keyboard; instead, the user can touch the surface with fingers. As the name suggests, multiple many users can work simultaneously as the table 'remembers' who 'possesses which objects. The authors suggest that this technology could be very appropriate at a tactical level in commando centers, where disaster managers have to discuss steps in managing emergency situations.

The chapters in this book are aimed at researchers, practitioners, and students who apply remote sensing technology in monitoring of hazards and managing of disasters. The book itself is the result of a collaborative effort involving 37 researchers located in 8 countries.

Acknowledgements

The editors express their sincere gratitude to IRSS, India, the RGI-029 project 'Geographical Data Infrastructure for Disaster Management', the Netherlands and Delft University of Technology, the Netherlands for making this book possible.

Previous book volumes in the series of Gi4DM

Zlatanova, S. and J.Li, 2008, Geospatial Information Technology for Emergency Response, 2008, ISBN 13: 978-0-415-42247-5 (hbk), ISBN 13: 987-0-203-92881-3 (ebook), Taylor & Francis Group, London, UK, 381 p. (ISPRS book series Vol. 6)

Li, Zlatanova, Fabbri, 2007, Geomatics Solutions for Disaster Management, 2007, ISBN 10 3-540-72106-1 Springer, Berlin, Heidelberg, New York, 444 p.

Van Oosterom, P., S. Zlatanova and E. Fendel, 2005, Geo-information for Disaster Management, 2005, ISBN 3-540-24988-5, Springer, Berlin, Heidelberg, New York, 1434 p.

Part 1: Use of Geo-Information technology in large disasters

1. Geoinformation-Based Response to the 27 May Indonesia Earthquake – an Initial Assessment

Norman Kerle and Barandi Widartono

Abstract

A devastating earthquake occurred on 27 May 2006 in the Yogyakarta area in Indonesia. Response activities began immediately, and included extensive ground-based mapping by Indonesian entities, as well as an activation of the International Charter "Space and Major Disasters", which led to the rapid production of image based damage maps and other assistance. The aim of this paper is to assess the Geoinformation that became available and was used in the aftermath of the disaster. It shows that some of the map products, largely because of lack of field data and communication with forces in the disaster area, were not as effective as they could have been. It further provides a preliminary quality assessment of those damage maps, using data from a house-by-house damage assessment. Disaster response and data processing are still ongoing, and further analysis will be required to determine how the use of Geoinformatics, and the utility of international assistance based on Charter products in particular, can be improved.

1.1 Introduction

At 05:54 AM local time on 27 May, 2006, a magnitude 6.3 earthquake struck eastern Java in Indonesia. With an epicenter approximately 20 km SSE of Yogyakarta near the densely populated Bantul district, close to 6,000 people died and an estimated 154,000 houses were destroyed. Despite frequent geophysical disasters in Indonesia, the affected area had not experienced an earthquake of comparable magnitude in over 100 years, and was thus ill prepared. Simple brick buildings, the principal housing type in the affected area, could not withstand the motion and readily collapsed. Despite the time of the earthquake, very early in the day on a Saturday, many people were already busy outside their homes, limiting the loss of life in one of the most densely populated areas Indonesia with >1,600 people per km^2 (BAPPENAS, 2006).

1.1.1 The earthquake event

The earthquake occurred early on 27 May, at a shallow depth of approximately 10 km, and 20 km SSE of Yogyakarta (USGS, 2006), although epicenter coordinates indicated on various maps, as well as the hypo center depth, have varied substantially. While only limited damage occurred in Yogyakarta itself, the district of Bantul to the south suffered most, with additional substantial damage in the Klaten district to the NE. Preliminary estimates of damage exceed 3 billion US$, over 50% of which attributed to housing damage (BAPPENAS, 2006). The earthquake began just weeks after sustained strong eruptions at Merapi volcano, some 45 km to the North, prompting speculations of a connection. However, the movement followed a previously identified NE-SW-trending fault, and, while Merapi's activity may have played a role, resulted primarily from the Australian plate subducting beneath the Sunda plate at a rate of 6 cm per year (USGS, 2006). Figure 1 gives an overview of the affected area and seismic intensities caused by the earthquake. Note that the epicenter (star) as determined by the USGS is more than 10 km away from the fault line (hatched line) identified by Indonesian scientists as the source of the earthquake.

1.2. Immediate response activities

The disaster led to the immediate mobilisation of a variety of response activities. An USAID/OFTA team arrived on the same day, followed quickly by other organisations over the next few days. Within Indonesia, the established disaster management hierarchy was activated, comprising of Bakornas, Satkorlak and Satlak for the national, provincial and district levels, respectively. The latter coordinated the local work, predominantly in the Bantul and Klaten districts. In addition to these efforts, the Ministry of Public Works carried out a rapid damage assessment for 300 selected public buildings to assess structural integrity, followed by an extensive house-to-house mapping campaign by the geography department of Gadjah Mada University (UGM) in Yogyakarta. About 100 staff and students carried out several ground mapping projects. After an initial survey of the emergency, a rapid building damage survey was initiated, followed by a more detailed one. Data entry and processing are still ongoing, and the large amount of data promises to be valuable in assessing the accuracy and potential limitations of purely image-based damage maps. A preliminary assessment of the parts of the UNOSAT damage maps is provided below. It is based on data for one of 8 mapped districts, Imogiri, for which alone over 14,000 houses were individually assessed for damage.

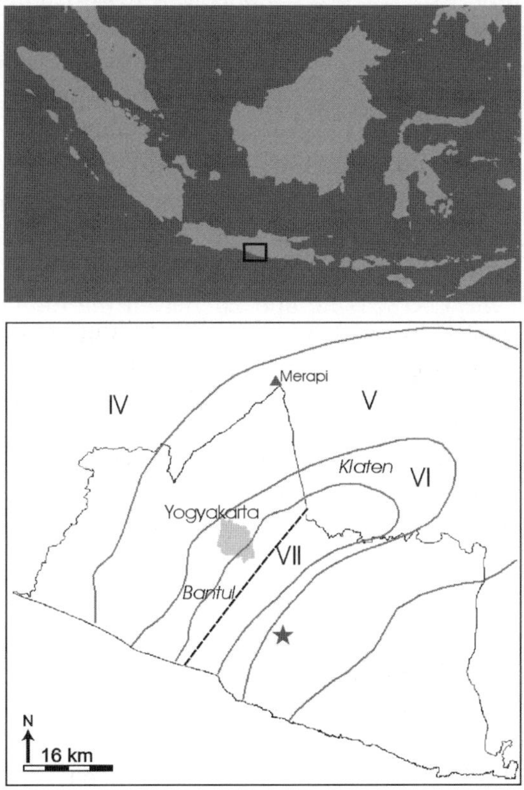

Fig. 1. Map of the disaster area, with seismic intensity zones and the mainly affected districts of Bantul and Klaten, as well as the active fault (hatched line) and epicenter (star) indicated.

The International Charter was also activated on the day of the earthquake by the German Foreign Ministry, and project management assigned to the German Space Agency, DLR. Due to favourable satellite positions and pointability, high resolution satellite data were acquired as early as 28 May, and again on 30 and 31 May, while medium resolution ASTER and SPOT images were taken on 30 May. In addition, Japans Daichi satellite passed over the area on 28 May, collecting images with the AVNIR (VNIR) and PALSAR (radar) sensors. A pre-disaster Ikonos image was also acquired on 9 May 2006, while 2 Quickbird scenes were taken in July 2003, all adding to a substantial array of data. Figure 2 shows the images and ground data that were obtained, and the damage maps produced.

Despite the abundance of images, however, there were also problems. In particular (i) the immediate availability only of lower resolution quick-looks, (ii) distribution by different vendors and organisations at different prices, (iii) availability of the actual satellite data, as opposed to quick-look or jpg-images, only to selected users, and (iv) frequent distribution in smaller tiles led to difficulties. For example, a largely cloud-free Ikonos image covering much of the affected area was acquired on 28 May and was available from CRISP in Singapore, the Asian distributor for Ikonos data. However, LAPAN, the Indonesian mapping agency only received and distributed parts of it (several tiles missing; Fig. 3), while the DLR carried out a preliminary damage assessment only on two lower resolution samples posted on the CRISP site, and later used the cloudier Quickbird image that covered a smaller area. This was because the CRISP Ikonos images are nearly 3 times the price of Ikonos imagery sold through the European vendor, exceeding the budget made available by RESPOND, the GMES service element for the use of Geoinformatics for humanitarian assistance (DLR, pers. comm.).

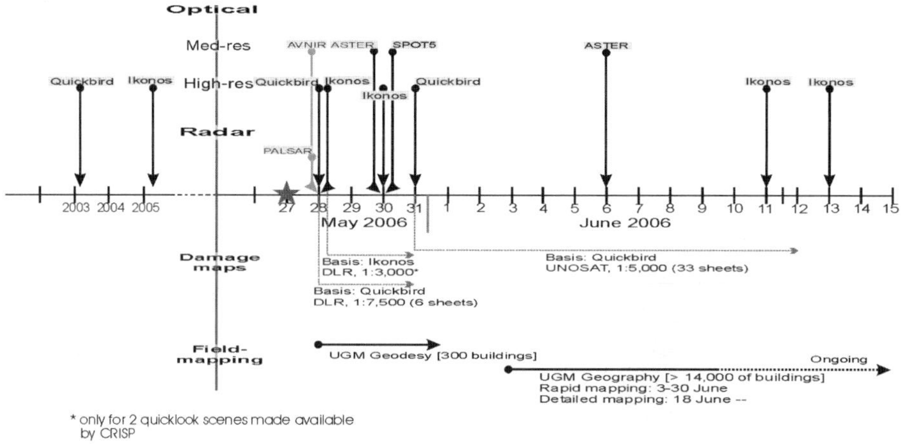

Fig. 2. Overview of high and medium resolution image data acquired before and after the 27 May earthquake. Damage maps based on those images, as well as field mapping carried out, are also indicated.

1.2 Damage Map Products

Image-based damage mapping is constrained by the spatial resolution of the available data and the average size of the destroyed objects to be mapped. If those objects are too small to be imaged individually, texture-based processing is required, by necessity resulting in lower accuracies (Kerle et al., 2008). Given the nature of the predominantly affected houses – small and distributed in clusters amidst heavy vegetation – high resolution data such as Quickbird or at most Ikonos are required. The price for such high detail, however, is not only high image cost, but also lower coverage. From Fig. 4 it is clear that, despite the seemingly extensive image database, only limited high resolution coverage of damaged areas was achieved, further hampered by clouds. Hence the substantial mapping efforts could only produce detailed damage assessment of parts of the area, a typical reality for such disasters, especially spatially extensive ones in tropical areas. Additionally, damage mapping as part of a Charter activation is carried out on a best-effort basis, typically without feedback from the field, constrained by the available or affordable data, as indicated above, and with very limited time. Therefore, focus on especially affected areas at the expense of comprehensive coverage is frequent.

Following the 27 May earthquake a large number of map types was produced by different entities, the majority of them based on field information. Reliefweb (www.reliefweb.int) lists some 50 maps for the earthquake area, primarily produced by UNOSAT, DLR, IFRC, MapAction, and OCHA. All of them are of high cartographic standard and optimised for large-scale printing, and reflect an overall increasing specialisation of different response organisations, each providing assistance according to its specific expertise and resources, and generating map products related to specific aspects of the disaster.

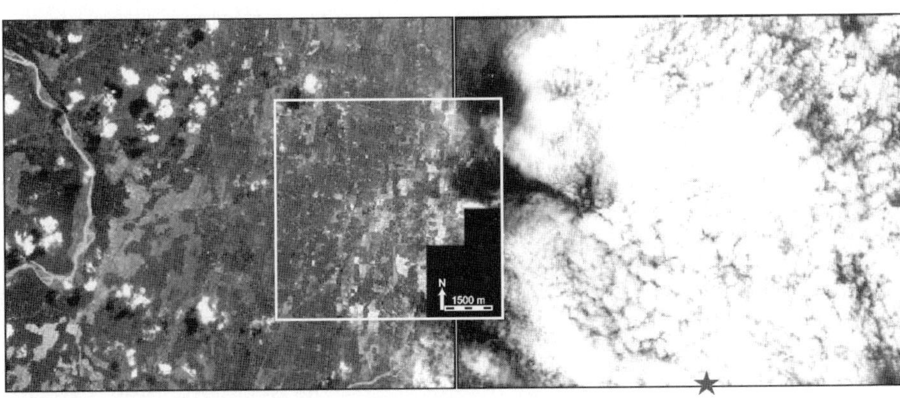

Fig. 3. Ikonos scenes of the disaster area acquired on 28 May, available through CRISP in Singapore (background), and actual scene provided to LAPAN in Indonesia, all color images reproduced here in grayscale. Star shows epicenter. For footprint of the smaller image see Fig. 4.

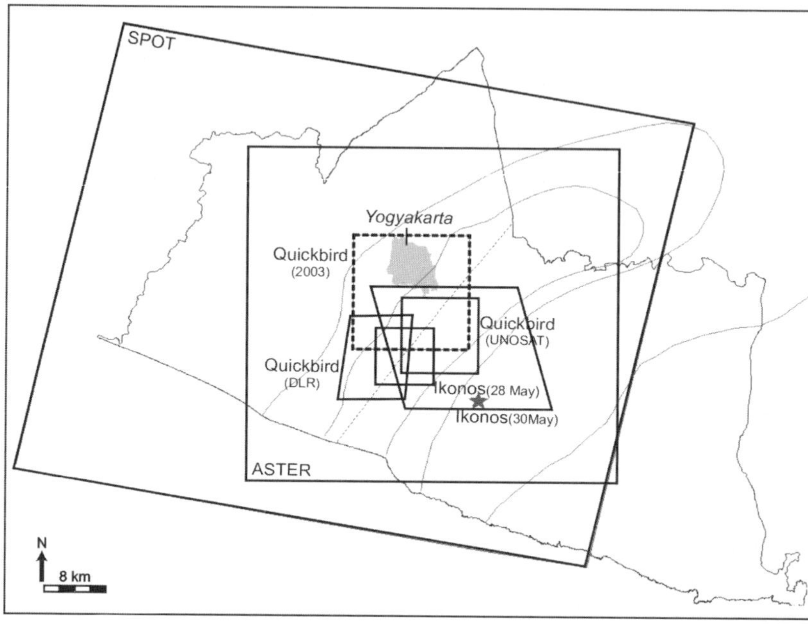

Fig. 4. Footprints of selected pre- (hatched line) and post-disaster (solid lines) satellite images that covered parts of the affected area.

For Indonesia, only DLR and UNOSAT produced image-based damage maps, which include (i) overviews, typically annotated with auxiliary information such as coordinate grids, place names, major roads, etc., (ii) pre- and post- event comparisons with or without further analysis, and (iii) stand-alone damage maps. Each of those categories can contain further sub-types (Fig. 5). All high-resolution versions can be found on Reliefweb, RESPOND (www.respond-int.org), DLR-ZKI (www.zli.dlr.de) and UNOSAT (www.unosat.org).

Maps created after a disaster are aimed at different purposes. They include simple reference maps that facilitate orientation and navigation, status-quo maps of different disaster-related aspects such as landslides or building damage, auxiliary maps showing population densities, utility lines, ethnic distributions, etc., and maps as a planning basis. Those may show possible shelter locations, or areas safe for reconstruction. Depending on their information content, the maps are needed by different users and at different times, and are ideally produced with a specific user group in mind. Especially maps prepared in the first few days after an event, however, are typically prepared far away from, and without a direct communication link to, the disaster area. This results in the risk that maps are prepared without awareness of the specific information needs of, in particular, national response forces in the affected country. Similarly, one of the limitations of the Charter is limited knowledge within potential beneficiary countries on what the Charter may provide, and when. In Indonesia this led to a situation where local institutions carried out their own damage mapping, albeit ground-based but with satellite image support. This was approached from 2 sides. First, as stated above, a limited assessment of public buildings was carried out by the Ministry of Public Works, while UGM organised a more extensive house-by-house mapping campaign that lasted over one month. In addition, bottom-up reporting worked well, where representatives from some 1,200 villages reported damage numbers to district officials, and from there further to the district. This took approximately 2 weeks to be done, longer than the initial image-based maps, but also resulted in more reliable data.

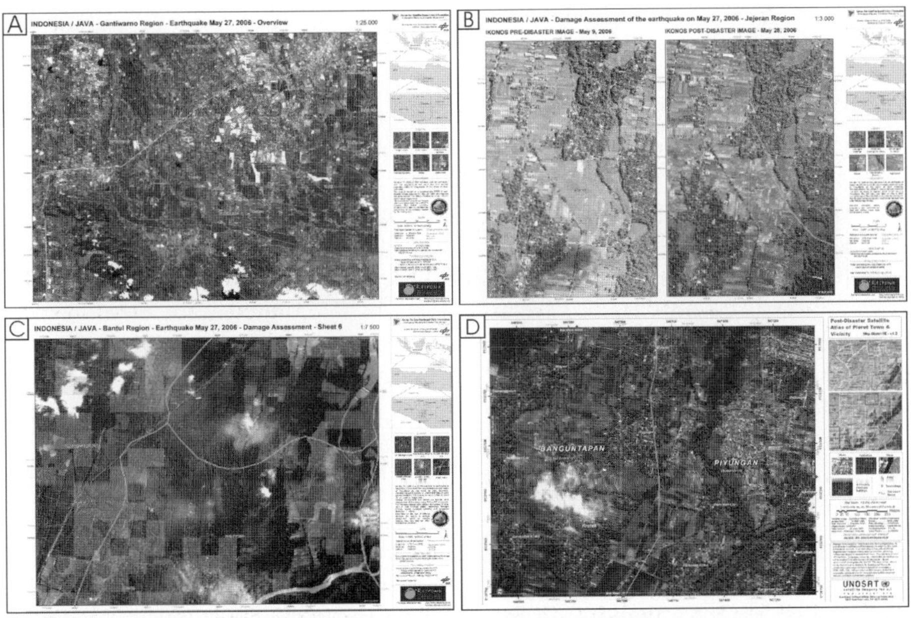

Fig. 5. Illustration of different image-based map types. A: overview map based on SPOT5 data with auxiliary data overlaid; B: damage map based on pre- and post disaster Ikonos data; C: 250m grid damage map based on Quickbird image; and D: more detailed damage mapping on the same Quickbird image (source: A-C DLR, D UNOSAT). All maps are originally on color.

This is particularly so since only a coarse damage map based on 250m grid cell was prepared by DLR at first (published on 31 May), while map results from a more detailed assessment in the area further East that was located in the highest intensity zone, were only published by UNOSAT on12 June, the same day as the map based on village reports (prepared by OCHA, available on Reliefweb).

1.3 The Response Chain and the Weakest Link

Geoinformation has a tremendous potential to facilitate all aspects of disaster management, including response. The general approach is illustrated in Fig. 6, which shows the main elements of the chain, as well as constraining factors.

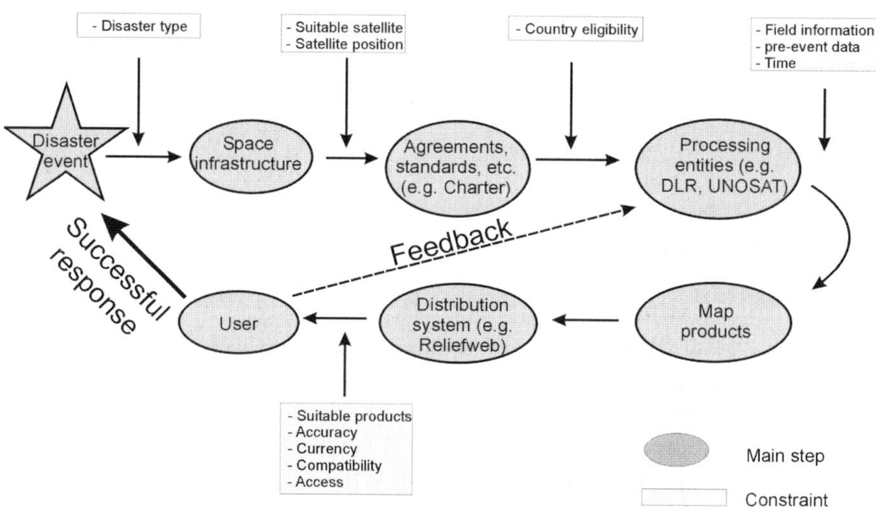

Fig. 6. The Geoinformation-based disaster response chain.

To what extent image-based information can aid in an emergency situation depends primarily on the disaster type (Kerle et al., 2008; Zhang and Kerle, 2008). Suitable sensors are then further constrained by their actual availability, a function of position and pointing capabilities. Civilian data acquisition and distribution is still a relatively slow process, hence agreements such as the Charter have been establish to speed up the process. The acquired data are made available to reference imagery, as well as more analysis time. The map products thus produced can be highly variable in quality. SERTIT, for example, when in charge of a Charter activation, attempts to produce damage maps within 12 hours of image acquisition (SERTIT, pers. comm.). Current distribution of map products is almost entirely internet-based, which allows rapid and lossless global circulation and easy reproduction. Sites dedicated to disaster management support, such as Alertnet or Reliefweb, or sites by the organisations involved in the Charter are very effective in making this information available. The most significant constraint in the chain is the link to the user. Only if map products suit the need of the user, are timely and current, accurate, compatible with existing data, and can be readily accessed, is there a potential to aid in the emergency response.

1.4 Preliminary Accuracy Assessment of Charter Damage Maps

As shown above, the damage maps prepared by DLR and UNOSAT, with the exception of 2 small Ikonos scenes, were based on cloudier or older (31 May) Quickbird data, although its higher spatial resolution also allows more detailed mapping than would have been possible with initial Ikonos data. So far only a small amount of the ground-based damage mapping data are available, which only partly overlap with the UNOSAT damage maps. Of the approximately 14,000 mapped houses in the Imogiri district (dark dots in Fig. 8), some 6500 were completely or heavily damaged, though of those less than 2000 fell into the area covered by the Quickbird image (Figs. 7 and 8). Figure 7 shows part of the 31 May Quickbird used by UNOSAT. *B* shows heavily or completely damaged areas as indicated in those maps, as well as actually heavily damaged or destroyed (squares) or moderately affected (circles) houses as mapped by the UGM campaign. *C* shows a close-up of the image, indicating the how well complete damage is distinguishable in Quickbird data. Parts of the image, in particular in well-exposed damage clusters, were accurately mapped (*D*, location shown in *B*), while more isolated houses in denser vegetation, or structures with less than total damage were harder to identify (*E*). Many more data are expected for the other areas mapped rapidly or in detail (outlines in thick solid and thinner hatched lines, respectively, Fig. 8), and will be analysed in the coming months.

1.5 Conclusions and Preliminary Lessons

Following the 27 May earthquake, extensive ground-based damage mapping was initiated by UGM, while the Charter activation made rapidly acquired high-resolution satellite data available to DLR and UNOSAT. This particular disaster was unusual insofar as a large number of multi-source data became quickly available. Conversely, it was an event quite comparable to other recent disasters such as the earthquakes in Pakistan, Gujarat or Bam, where damage was severe and widespread, confusion reigned initially, and the detailed picture of the destruction only gradually emerged.

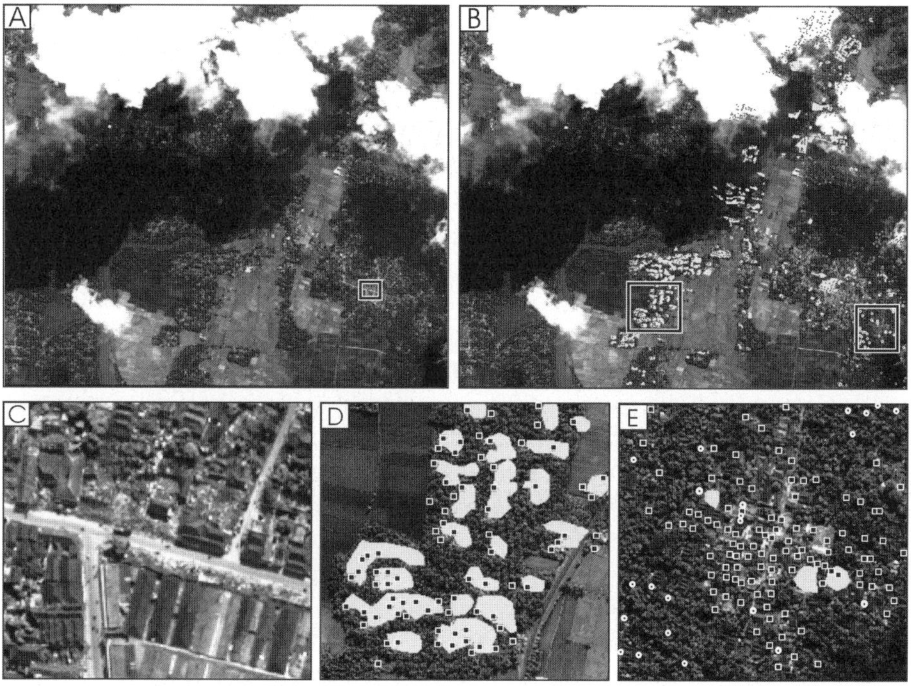

Fig. 7. Part of the Quickbird image of 31 May that was used by UNOSAT for damage mapping (A). See text for details.

Although the Charter was activated and damage maps were produced, knowledge from the field is never or scarcely incorporated in this process. In the Indonesia case it appears that local authorities did not know what products to expect, or when, and hence resorted to their own mapping, benefiting from the proximity of a well set-up university. Here the existing administrative hierarchy also facilitated the damage reporting from villages up to the province level. As illustrated in Fig. 6, the weakest link determines the use of damage maps and other information made available through the Charter. Here this was the missing link to the user, in particular the Indonesians. While international aid organisations are more likely to have good knowledge and experience concerning Charter products – their value as well as limitations – the authorities in the affected country typically do not. Hence the quality and speed of delivery of such products becomes less relevant, and a good way to improve the impact of such maps is to spread awareness of likely assistance, and to involve local knowledge from the very beginning.

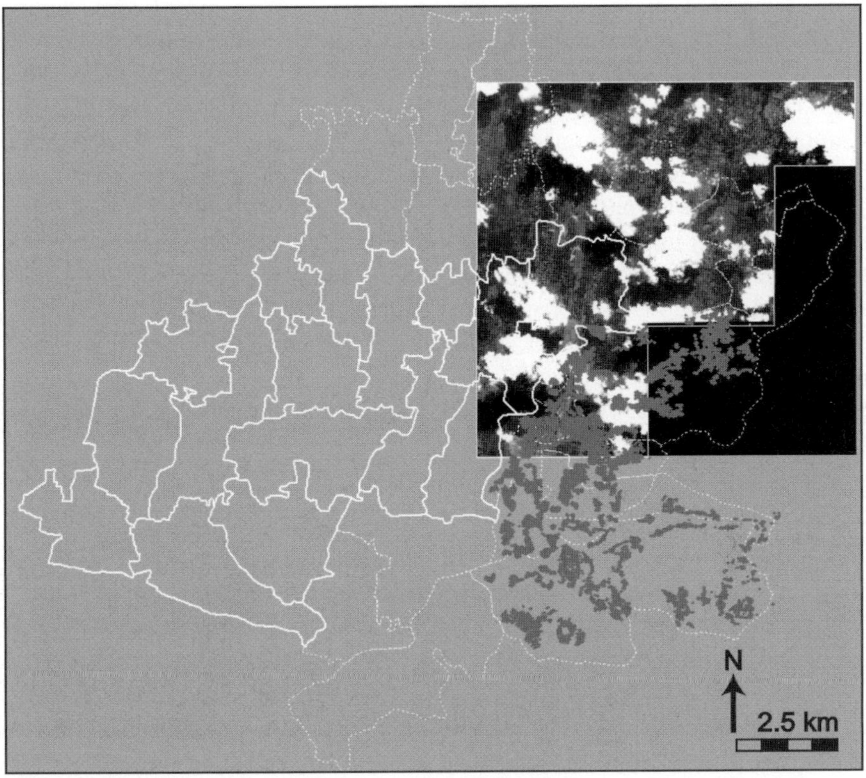

Fig. 8. Quickbird image of 31 May used by UNOSAT for damage mapping, and areas covered by the UGM rapid and detailed (marked by thicker solid and thinner hatched white lines, respectively) building damage survey. The 14,000 data points for one of the 8 districts, Imogiri, are shown as dark dots.

It is also apparent that one-fits-all damage maps are produced. While great effort and best intentions go into this process, the lack of user needs consideration undermines the eventual utility of the maps, a point that needs to be investigated in detail. Similarly, the maps frequently lack information on the underlying assumptions of the analysis, and the expected accuracy and hence use of the map. DLR, for example, produced the aforementioned initial maps based on 250 m grids, illustrated in Fig. 9. The maps show different damage levels and state that deviations from ground-based assessment are possible. Unfortunately the results from the UGM field mapping campaign for the area covered by the DLR maps are not yet available, precluding at quantitative quality assessment at this point. However, parts of the image clearly show significant damage (Fig. 9, middle), yet are not assigned to any of the 3 damage classes. A detailed assessment with house-by-house data will likely be useful better to understand mapping limitations, but also to evaluate how useful it is to produce maps with minimal mapping time with an accuracy that cannot be assessed. Work is currently being carried out on other issues as well, such as the potential of small format aerial photography (SFAP) for rapid damage assessment, and to identify the optimal scale vs. coverage parameters and limitations of that approach. For the disaster area it is also important to understand better the performance of different building types in relation to seismic intensity and location, in terms of topography and soil type, and to understand better seismic risk in the area. Rebuilding has already begun in the area, hence such work need to be carried out soon if it is to contribute to improved resilience.

Fig. 9. Part of a damage map using a 250 m grid (damaged areas outlined in white), prepared by DLR based on 28 May Quickbird data (left), and the same area covered by the 31 May Quickbird image, illustrating potential mapping inaccuracies. For example, the area shown in the middle (small boxes show location) shows heavy damage, which is not marked as such in the DLR map.

As mentioned before, the Japanese PALSAR sensor acquired data of the disaster area on 28 May. The value of such information has been repeatedly investigated, though with mixed results (Arciniegas et al., 2007, Fieldings et al., 2005). PALSAR's specific promise, in addition to cloud-penetration, lies in its longer wavelength L-band (23.5 cm) as opposed to previously available C-band (3-5 cm), which may improve its performance in vegetated areas. Unfortunately, the Daichi satellite is still in its validation phase, hence the data were not available for the 27 May disaster response, but a detailed assessment of its interferometric potential for damage mapping needs to be explored.

Acknowledgments

We thank colleagues at Gadjah Mada University, in particular Drs. Hartono and Sudibyakto, and Heri Sutanta, as well as Hiroshi Watanabe at ERSDAC in Japan.

References

Arciniegas G, Bijker W, Kerle N, Tolpekin VA (2007) Coherence- and amplitude-based analysis of seismogenic damage in Bam, Iran, using Envisat ASAR data. IEEE Transactions on Geoscience and Remote Sensing, Special issue "Remote sensing for major disaster prevention, monitoring and assessment": 45

BAPPENAS, Government of Yogyakarta, and international partners (2006) Preliminary damage and loss assessment: Yogyakarta and Central Java natural disaster. Jakarta, Indonesia, p. 17

CRED (2006) EM-DAT: The OFDA/CRED International Disaster Database, www.em-dat.net

Fielding EJ, Talebian M, Rosen PA, Nazari H, Jackson JA, Ghorashi M, Walker R (2005) Surface ruptures and building damage of the 2003 Bam, Iran, earthquake mapped by satellite synthetic aperture radar interferometric correlation. Journal of Geophysical Research-Solid Earth: 110

Kerle N, Heuel S, Pfeifer N (2008) Real-time data collection and information generation using airborne sensors, In: Zlatanova S, Li, J. (eds) Geospatil information Technology for Emergency Response, ISPRS Book Series Vol. 6, Taylor & Francis, pp. 43-74

USGS, 2006. Earthquake Hazards Program, http://earthquake.usgs.gov/

Zhang Y, Kerle N (2008) Satellite remote sensing for near-real time data collection. In: Zlatanova S, Li, J. (eds) Geospatil information Technology for Emergency Response. ISPRS Book Series Vol. 6, Taylor & Francis, pp.75-119

2. The Application of Geo-Technologies after Hurricane Katrina

Henrike Brecht

Abstract

Geographic Information (GI) Science accelerated and enhanced decision-making in the emergency response after Hurricane Katrina. Since most of the information needed in disaster management has a spatial dimension, geo-technology is a vital source for streamlining response activities. The rapid-response situation after a disaster, however, exposes new challenges in the use of geo-technologies. In order to move forward and enhance the GI applications in disaster response, this paper pinpoints the bottlenecks and highlights the successes of the use of geo-technology after Hurricane Katrina. Challenges and accomplishments in the response to the storm are analyzed and lessons learned are documented in the five areas of management, technology infrastructure, data, workflows, and map products. One of the explanations for the experienced bottlenecks is traced to difficulties with regard to timely data access and dissemination; one of the successful practices was the intensive integration of web-based tools.

2.1 Introduction

While mainstreaming geo-information in disaster response is becoming increasingly recognized as a key factor for successful emergency management, systematic knowledge about the benefits and bottlenecks of geo-technologies in the response phase is still in fledging stages. In the complex, dynamic, and time-sensitive disaster response situation of Hurricane Katrina, geo-information enhanced decision-making and effectively supported the response but it did not reach its full potential. The overwhelming complexity of the disaster exposed challenges and highlighted good practices. Hurricane Katrina affected an area of nearly the size of the United Kingdom (230,000 square km), it killed more than 1,700 people, and the total cost of damage is estimated at more than $200 billion dollars. The destruction, which has affected primarily the coastal regions of Louisiana,

Mississippi, and Alabama, was caused by high-speed winds, storm surge flooding in coastal areas, and in New Orleans also by levee failures.

Information management is a crucial component of emergency response. The ability of emergency officials to access information in an accurate and timely manner maximizes the success of the efforts. Since most of the information used in disaster management has a geographic dimension (Bruzewicz 2003), geo-technologies have a large capacity to contribute to emergency management. The capabilities of geo-technologies to capture, store, analyze, and visualize spatial data in emergency management have been documented in the literature (Cutter 2003; Zlatanova 2006; Carrara and Guzzetti 1996). Paradoxically, in praxis the convergence of the two fields of geo-information and emergency management is only rudimentary developed and little work has been undertaken to enhance the integration.

What were the bottlenecks of using geo-information in the response phase of Hurricane Katrina? Which mapping services were requested frequently? Which workflow procedures streamlined the mapping support? What were the best practices?

In the following these questions are addressed focusing on five areas:

- managerial lessons with regard to information flows and staffing issues;
- the perfidies of technology infrastructures in an emergency situation;
- important datasets and best practices of data documentation and access;
- workflows that streamlined the mapping response;
- the "stars" of the mapping products, which were requested or needed the most.

2.2 Lessons Learned

The knowledge about best practices was gained from the experience of GI responders. Input was gathered mainly during the Louisiana Remote Sensing and GIS Workshop (LARSGIS) in Baton Rouge, Louisiana, in April 2006 in which practitioners from the coastal southeastern United States presented and discussed their experiences of using geo-technologies after Hurricane Katrina. The author's own experience in the Emergency Operations Center (EOC) of Baton Rouge after the storm also influenced this paper.

2.2.1 Managerial Lessons

Improving Information Flows

Large amounts of data were acquired and processed after Hurricane Katrina. In the immediate aftermath of the disaster, governmental agencies and private geo-technology companies, realizing the extent of the damage and the gravity of the situation, supported the relief efforts by contributing data. Numerous sets of aerial photographs were taken and distributed to assess flooding and damage, private companies donated satellite images, data, and hardware, and new data layers concerning emergency shelters or power outages were created. Public agencies released and shared existing but previously undisclosed data layers. The usual obstructive administrative barriers caused by competition and conflicts between divisions were abrogated, and instead ad-hoc alliances were built to support the common goal of saving lives and containing the devastation. Data streamed in quickly, resulting in the availability of a multitude of new data layers. The dissemination of the data to the appropriate parties at the desired locations in a timely manner, and in a useful format may have been the biggest challenge for the GI response community. Agencies were not always aware which information was available or where to find certain data. Due to miscommunication, excessive workloads, and general distress, information was distributed only to a limited extent and did not always reach the first responder crews or county governments in remote areas that were in crucial need of this information.

Information flows and structures between the different actors must be identified before the disaster. One possible strategy is to appoint a central data authority that collects and disseminates information, a solution that is effective but difficult to realize due to political and economic reasons. Spatial data infrastructures and web-based solutions have proven to enhance information flows and data accessibility. These tools should to be established before the disaster strikes.

Establishing Geo-Technologies as an Integral Resource

Mapping support often evolved as an ad-hoc component after the storm being triggered by a high demand for maps and geo-information. Impromptu volunteers were engaged or geo-information companies were hired on the spot. Emergency preparedness units need to recognize geo-technology as crucial part of disaster management and incorporate it accordingly into their planning. It is the task of the GI community to increase the awareness of emergency managers towards the value of spatial technology. During the emergency knowledge gaps became apparent on both sides: governmental emergency staff was unclear about the potential of geo-technologies and the use of maps and the GI community was not informed about governmental disaster plans and strategies. Both parties have to gain an increased understanding of each other's duties and capabilities. Communication and training platforms are means to enhance awareness.

Building Partnerships
Formal and informal partnerships between GI professionals that were established before the disaster proved to be essential in the disaster response. Relationships facilitate coordination and thus the flow of information. One way to strengthen collaboration is the establishment of a workgroup of GI-skilled personnel in governmental agencies, universities, and private industries. Regular meetings foster networks and enable the exchange of news about available data and technologies.

Identifying Staff
GI responders were confronted with many requests for maps and an understaffing in the EOCs. It proved valuable to call on the support of GI colleagues. Volunteers played an important role in the response to Katrina, and it is recommended to integrate them into emergency planning. Staff to support operations during an emergency needs to be identified beforehand. If a disaster occurs, a call-up of pre-defined GI-skilled personnel should be initiated to assemble teams. The response teams should include staff from different governmental departments and from academia, assembling specialists from the different fields in geo-technology, such as remote sensing, programming, databases, and GIS. It is helpful to allocate staff to certain responsibilities pertaining to data collection, logistics, technical support, mapping, distribution, and operational management. Specific staffing challenges are caused by the 24 hours per day, seven days per week operations which require a high staff rotation. For the rotation not to affect efficiency, detailed documentation of requests, actions, files, and file locations are necessary.

2.2.2 Technology Infrastructure Lessons

Ensuring Hardware Resources
The EOCs were not or only rudimentary equipped for geo-technologies prior to Hurricane Katrina. Computers, plotters, printers, and other supplies had to be identified and installed after the storm. Difficulties occurred with regard to finding space in the EOCs not only for large hardware devices and storage systems for hard-copy maps, but also for laptops and workstations. Mapping teams should establish sources and localities of all

necessary hardware beforehand and explain their special demands so that physical space in the EOCs can be allocated. In the response to Hurricane Katrina, innovative solutions were found, such as the one from a mapping team in Mississippi that remodeled a bus into office space and equipped it with workstations and printers.

Securing Continuity of Operations

Useful datasets were stored on computers that flooded or that were left behind in the evacuation. Data back-ups at multiple secure locations and mobility of hard- and software are to be established to enable continuous operations under emergency conditions and to avoid loss of data. Data accessibility was not only hampered by disrupted networks and flooded computers but also by logistical issues. In one case, important files were passport-protected and the responsible administrator could not be reached.

Preparing for Power and Network Disruptions

Power, network, and internet outages were frequently encountered. Ideally, alternative power supply solutions are identified beforehand, including generators and uninterruptible power supplies (UPS) which are battery backups that can be added to hardware devices to avoid data losses during power disruptions. Since it is not advisable to rely on network connectivity, sufficient data sharing devices are necessary for an efficient response. Moreover, regular back-up mechanisms proved to be valuable.

Administering Networks

Not only GI skills were vital for successful operations but GI staff installed intermittent network routers, virtual private networks and other network connections. Ideally, a network administrator is appointed who is in charge of connectivity issues.

2.2.3 Data Lessons

Acquiring Relevant Data

Base datasets, for example about pumping stations, utility networks, and power plants, were not always readily available. Especially for rural areas, geo-information was scarce. Information that proved to be of focal interest during the emergency can be divided into two categories: information that should to be collected before the disaster and information that is to be collected after the disaster.

Before the Disaster

Datasets that were vital during the response and that can be acquired before the disaster include but are not limited to:

Pumping stations	Hazardous materials
Street maps	Building footprints
Elevation models	Helicopter landing places
Points of interest	Special needs population
Fire stations	Evacuation routes
Cadastral data	Population densities
Medical centers	Day and night population
Geomorphology	Utility networks
Land use	Emergency resources
Power plants	Address dataset
Satellite imagery	

After the Disaster

Datasets that were frequently requested in the EOCs providing information on the extent of the catastrophe include but are not limited to:

Wind fields	Oil spills
Power outages	Flood depths
Debris estimates	Levee breaks
Daily dewatering	Road restoration
Power restoration	Emergency shelters
Flood fatalities	Flood extent
Satellite imagery	Fire outbreaks
Deceased victim locations	Points of dispensing
Restored power	Pollution
Crime scenes	Damage estimates

Sources need to be established for information that becomes available after the disaster. This can be accomplished with data sharing agreements, which should be set up prior to the emergency. These agreements determine which data will be provided by which organizations and who holds copyrights. For instance, uniform, useful, and complete image datasets were in high demand after Katrina. Therefore, contracts with companies providing aerial photography should be in place, specifying resolutions, area coverage, formats, geo-correction procedures, and accompanying metadata. Agreements need to include how often datasets will be updated since some of the mentioned data layers require daily updates. For instance, shelter locations opened rapidly in the immediate aftermath and then, after a few weeks, closed or moved. Information on flooded roads also needed daily updating, as did the locations of crime scenes.

Clarifying Copyrights

The clarification of data copyrights and privacy laws was time-consuming. It was difficult to reach those in charge to get permission for data dissemination because communication networks were interrupted, electronic address books were inaccessible due to flooded and left behind computers, and officials were dispersed because of the evacuation or not available during the weekend and at night. It is of advantage to negotiate data dissemination agreements, data sharing policies, and specifications of data custodianship before the disaster.

Collecting Metadata

After Hurricane Katrina, a multitude of datasets were disclosed and created rapidly. Maps showing the newly available information were requested, produced, and distributed in extremely short time spans. A central problem that arose from this incoming data stream and the stressful situation was that metadata tended to be neglected. However, crucial information is rendered unemployable if datasets are not properly documented. Moreover, metadata helps to maintain standards for data quality. Finally, missing metadata causes delays since valuable time is spent struggling to find out, for example, on which date an aerial photo set was taken and which area it covers. A metadata standard should be chosen that answers questions of data timeliness, source, accuracy, and coverage. Although metadata collection is time-consuming, GIS staff receiving data must be dedicated to metadata collection, ensuring that a predefined form is completed for all incoming datasets. A data manager should be assigned whose responsibilities include documenting metadata.

Organizing Data

In the response phase, geographic information must flow upstream and downstream between players in real-time. An effective means of accomplishing this dissemination of data is a spatial data infrastructure (SDI) which enables an efficient, reliable, and secure way for the search, exchange, and processing of relevant information. An SDI is a framework that subsumes a collection of geospatial data, technologies, networks, policies, institutional agreements, standards, and delivery mechanisms. Creating an infrastructure subsuming both general and emergency-related data with clearly laid out directory structures and logical names is critical for effective emergency response where many applications occur in real-time. The SDI datasets need to be updated continuously, and data integrity has to be maintained. The responsibility of data creation and maintenance for the SDI cannot lie with one individual organization; it must rather be a joint effort of many organizations.

2.2.4 Operational Lessons and Workflows

Avoiding Duplication of Efforts
Duplication occurred when maps, conveying identical information (e.g. damage levels, road flooding, or power outages), were created by several agencies. Coordination via the implementation of a map depository where central players submit and download maps is a possible solution to this duplication.

Tracking Requests
Keeping track of map requests was conventionally handled by means of paper files. A team from the Louisiana State University implemented an online tracking system that largely improved the paper system. This tracking system not only documented the actual request but also associated information including contact information of the client, file locations, and map products. The system allowed efficient communication with all members of the response team, which was particularly important due to the high staff turn-over and geographically distributed mapping operations. The documentation of file locations and templates was especially helpful for the preparation of the daily updates of certain maps such as road flooding and emergency shelters. Another feature of the system allowed personnel to be assigned to the various projects. Such a record system for requests, associated files, documents, staff, clients, and products proved to be useful and should be implemented before the disaster strikes.

Preparing Paper Maps
Despite increasing digitalization, paper maps were still essential for the response teams. A high demand of paper maps and only limited printing and plotting capacities caused delays in fulfilling requests and disseminating information. Base maps, especially street maps on different scales, can be prepared beforehand. Ensuring access to sufficient amounts of paper, printers, and plotters is crucial.

Creating Templates
Map templates were found to be useful in the response activities. In the case of daily updated information, consistent templates accelerated the creation of maps and facilitated the comparisons of changes. Predefined map templates containing many data layers, which are turned on and off according to specific needs, saved a considerable amount of time. The templates should be well documented and logically stored within the data structure.

Disseminating GIS Resources

It proved valuable to disseminate GI operations. While staff on site in the EOC took requests, promoted geo-technologies, offered solutions, and generated quick maps, staff in remote locations was able to create more sophisticated maps and provide analysis away from the whirl in the EOC. This approach also guaranteed access to hardware, especially printers, that were not backlogged as was often the case in the EOC.

Using Online Tools

Web-based tools such as Mapquest or Google Earth, were used intensively in the response operations. Not only the GI staff but many of the involved responding agencies and rescue workers applied especially Google Earth for their operations. Google Earth and Google Maps created satellite imagery overlays of the devastation in the affected region, which helped to understand the scope of the disaster. Single houses and addresses could be looked up in Google Earth, and a built-in transparency slider, which allowed to switch between before and after images, enabled to see if and how much damage a place experienced. The accuracy, ease of access and the ease of use at the time and point of need of these online tools that can be operated by non-GIS staff contributed to the wide usage of the tool. This experience highlighted the potential of a web-based community approach to disaster operations.

Promoting Geo-Information

Since rescue workers were often not versed in the potential of geo-information it was useful to have a GI staff member attend official EOC meetings to offer GI-based suggestions and solutions. Another way to convey the GI services to official was by means of fixing frequently requested maps on the walls or collecting them in a map book for display.

2.2.5 Map Products

Requests from Emergency Responders

The large majority of requests from emergency responders were related to street atlases and area overview maps. Commercial maps in stores sold out quickly or flooded and therefore, responders relied on the GI community. Great numbers of street maps were handed out after the storm. The acquisition of digital copies of city maps from commercial companies was helpful. Emergency responders who were not familiar with the area requested maps with photos of landmarks such as the New Orleans Superdome. Checkpoints, which were still standing after the storm, were included in the produced maps. This concept proved to be remarkably helpful for an orientation of the area especially since many street signs were flooded or destroyed. Another crucial orientation and communication means is a grid system for ground reference. After the disaster, some responders considered missing grids in maps as one of the central problems. It would be helpful if map books in compact sizes with the standardized US National Grid, street indices, landmarks, and elevation levels are produced before the disaster. Overlaying street and area maps with satellite inundation maps that outline the extent of the flood and the flood depth was another frequent request. For instance, by means of this data, rescue missions determined whether boats or high-water vehicles are used in a certain area. For the search and rescue operations, mapping of addresses and coordinates of victims was of major importance in the first days after the disaster. Ideally, this process would be automated. Coordinates from a mobile phone placing a 911 call could be tracked automatically and then transferred to handheld computers of emergency responders. Geo-technology can calculate the best routes for accessing victims' locations.

The Needs of the Public

Accurate and timely information for the public is necessary. The questions of if a house was damaged plagued the evacuees. Days after the disaster, in order to find out if and how deep a house was flooded, the evacuated population relied on photos from television and the Internet to recognize neighborhoods and the levels of flooding and destruction. The uncertainty added to stress and anxiety. This information could be conveyed using web mapping and aerial photographs taken after the disaster. Vector layers with flood depths and levels of wind damage can complement the information. A system could be established that allows people to enter the address of a building to find out water depths, damage levels, when and how they could travel to the building, and nearest emergency supply centers. Moreover, the public requires detailed knowledge about the assigned evacuation routes and the traffic circumstances, evacuation shelters, kitchens, health facilities, and other public services.

Requests from Government Officials

Government officials asked for maps with various contents, including shelter locations, deceased victim locations, power outages, water systems, maps of state-owned land, pumping locations, and others.

2.3 Conclusions

This paper identifies lessons learned from the application of geo-technologies in the response to Hurricane Katrina. The main challenges in the operations were not related to the often discussed literature themes such as interoperability and semantics but rather to trivial issues such backlogged printers, network disruptions, and missing metadata. Aloof from Virtual Reality applications, one of the most frequent tasks of GI personnel was to print street books and visualize search and rescue coordinates. Truly analytical GI applications going beyond simple map displays were sparse. Experience shows that the suitability and use of GI technology in the response phase differs from the planning phase because of the urgency, uncertainty, the magnitude of stakeholders, some of whom are unfamiliar with geo-information, and the real-time data needs.

Among the most useful GI products created were inundation maps and map books with landmarks, detailed elevation figures, and unified grid systems. Web-tools such as Google Earth proved to be helpful due to their relevancy, ease of use, and open access. There are reasons to believe that the field of geo-information for disaster management will eventually benefit from further developments of visual environments, semantic interpretations, and other current research topics. Most likely, the future use of geo-technologies will extend beyond mapping and move towards analytical processes. This will especially be the case when emergency managers gain knowledge in geo-information and its capabilities. Currently, however, improvements on the ground are necessary on basic levels. Other analyses of geo-information in disaster management (Kevany 2003; Zerger and Smith 2003; Curtis et al. 2005) report similar experiences, stressing the practical impediments of implementations. Existing knowledge about best practices need to be translated into action, programs, and relevant policies. To enhance the response, the often separate discourses of geo-technology on the one hand and emergency management on the other need to converge.

References

Bruzewicz AJ (2003) Remote sensing imagery for emergency management. In: Cutter SL, Richardson, DB, and Wilbanks TJ (eds.) Geographical Dimensions of Terrorism. Routledge, New York, pp 87–97

Carrara A, Guzzetti F (eds.) (1996) Geographical Information Systems in Assessing Natrual Hazards. Kluwer, Dordrecht

Curtis A, Mills JW, Blackburn JK, Pine JC (2006) Hurricane Katrina: GIS Response for a Major Metropolitan Area. Quick Response Report QR180. Natural Hazards Center, University of Colorado

Cutter SL (2003). GI Science, Disasters, and Emergency Management. Transactions in GIS 7: 439-445

Kevany MJ (2003) GIS in the World Trade Center attack – trial by fire. Computers, Environment and Urban Systems 27: 571-583

Zerger A, Smith DI (2003) Impediments to using GIS for real-time disaster decision support. Computers Environment and Urban Systems 27: 123-141

Zlatanova S, Van Oosterom P, Verbree E (2006) Geo-information Support in Management of Urban Disasters. Open House International 31: 62-69

3. Application of Remote Sensing for Damage Assessment of Coastal Ecosystems in India due to the December 2004 Tsunami

Shailesh Nayak and Anjali Bahuguna

Abstract

Tsunami struck the Indian Coast including the Andaman and Nicobar Islands and the mainland coast on December 26, 2004. The tsunami 'run up' have significantly affected the coastal ecosystems on the Andaman and Nicobar Islands. Its effect on the mainland coast was less pronounced. Satellite data along with few ground surveys were used to assess damage to various ecosystems. Pre- and post-tsunami satellite data, mainly RESOURCESAT-1 AWiFs were used for the preliminary assessment. IRS LISS III and LISS IV data were also used in few cases. The impact on major ecosystems, such as mangroves, coral reef, sandy beaches, mudflats, tidal inlets, saline areas, forest, etc. was studied. The damage to ecosystems was categorized in to two types, total loss and degradation of ecosystems.

Mangroves of the Andaman and Nicobar Islands constitute about 18 per cent mangroves of the country. It was observed that the mangroves facing sea have undergone severe destruction. The mangroves have been eroded and trees have been uprooted. The mangroves along the creeks have been less damaged. Many mangrove areas remained submerged for a long time and thus destroyed. In the Andaman and Nicobar Islands, about 4700 ha mangroves were totally lost, while 10800 ha were damaged.

The giant tsunami waves smashed and crushed the reefs, while the backwash deposited the debris on the reef flat. The deposition of material killed the live coral colonies. It has been estimated that the Andaman and Nicobar Group of Islands lost reefs of about 50000 ha. About 11300 ha of reefs on the Andaman and Nicobar Islands were covered by sand, mud or detritus. The reefs of the Gulf of Mannar (south-east coast of India) have been affected marginally.

Most beaches have been affected on Islands and on the mainland coast. Few beaches have been washed away. On many beaches, sand has been deposited on the backshore. The nesting beaches for the endangered leatherback turtles that visit the Great Nicobar Island have been seriously damaged. On the mainland, sediments got accumulated on backshore in all tsunami-affected area. The deposition of heavy minerals, mainly ilmenite and garnet, were seen along the parts of the south-east and south-west Indian coasts. At few places, barrier islands/beaches were breached, on the south-west coast while the sand dunes were affected on the south-east coast. The tidal inlets on the mainland coast have either opened up or closed. In some cases, location of tidal inlets has also shifted. The lagoons have been contaminated by high saline water. Large tracts of land have become saline in the Andaman and Nicobar islands. About 25000 ha have been affected by salinity. Significant loss of forest, about 12000 ha was noticed on the Nicobar Islands.

The damage to ecosystem is critical as it directly affects livelihood of coastal communities. The enormous damage to the two vital ecosystems, especially the coral reef and mangroves, has affected the fishery resource in the area. Apart from some of the important species such as saltwater crocodiles, leather-back turtles and Megapode birds have been adversely affected.

3.1 Introduction

Tsunami waves struck the Indian coast on 26th Dec 2004, affecting the Andaman and Nicobar Group of Islands (6^0 45' – 13^0 14' N and 92^0 12' – 93^0 57' E), south-east and south-west Indian coasts. A tsunami is not a single wave but a series of traveling ocean waves generated by the geological changes near or below the ocean floor. The recent tsunami was set off as a result of a massive earthquake in the region registering a magnitude of 9.3, with its epicenter under the sea (more than 8-9 km below the sea bed), off the northern tip of the Indonesian archipelago near Aceh. The tsunami waves traveled as far as 6400 km from the epicenter. The tsunami moves rapidly across the ocean (900 km/h) and takes the form of destructive high waves along shallow coastal waters (10 m high and a speed of 40 km/h). The tsunami 'run-up' (a measure of the height of water observed onshore above mean sea level) travel fast and much farther inland than the normal waves.

Evidence from other tsunamis or severe storm events suggest that there are differences in the environmental impacts and that they could be caused by differences in coastal geomorphology (off-shore and on-shore), the presence or absence of off-shore coral reefs, the width of the mangrove belt, the maturity of the mangroves, and whether the mangrove is on a small island, large island or mainland coastline (IUCN 2005, CRI/ISRS 2005).

Satellite data was used to rapidly assess the damage caused by tsunami waves, particularly on the coastal wetlands of the affected areas. The Andaman and Nicobar Islands bore the maximum brunt as they are located closest to the epicenter and direct path of the tsunami. This paper describes the damage to the coastal ecosystems due to tsunami mainly using remote sensing data on the Indian coast.

3.2 Data used

Pre-tsunami and post-tsunami data of the Indian Remote Sensing Satellite-P6 (RESOURCESAT) AWiFS was primarily used to assess the impact. The details of the data are given below (Table 1). LISS III data was used wherever available. The specifications of the IRS P6 AWiFS and LISS III sensors are given in Table 2.

Table 1 Satellite data used for assessing impact of Tsunami waves

Area Coverage	Sensor (IRS P6)	Data available Pre-tsunami	Post-tsunami
Andaman & Nicobar I	AWiFS		4 Jan 05, 28 Jan 05*
		21 Dec 04	7 Feb 05
		16 Dec 04	2 Feb 05*
		22 Nov 04	30 Dec 04*
		17 Nov 04	8 Jan 05*
		29 Oct 04	18 Jan 05*
			7 Feb 05
			28 Jan 05
			16 Feb 05
	L3		30 Dec 04
			4 Jan 05
			30 Dec 04
South-east coast	AWiFS	23 Dec 04	28 Dec 04
			6 Jan 05
	L3		6 Jan 05
			7 Jan 05
South-west coast	AWiFS		27 Dec 04
		26 Dec 04	5 Jan 05

Table 2 Specifications of Sensor on board IRS-P6

S. No	IRS P6	Spatial resolution	Spectral resolution (μm)	Quantization
1	AWiFS sensor	56 m	Band 1 (0.52-0.59) Band 2 (0.62-0.68) Band 3 (0.77-0.86) Band 4 (1.55-1.7)	10 bits
2	LISS III sensor	23.5 m	Band 1 (0.52-0.59) Band 2 (0.62-0.68) Band 3 (0.77-0.86) Band 4 (1.55-1.7)	7 bits

3.3 Methodology

3.3.1 Satellite Data Interpretation

The satellite sub-images were extracted from all the data sets and subjected to radiometric and geometric corrections prior to unsupervised classification. Unsupervised classification was used to get the classification based on natural clustering. The Iterative Self-Organizing Data Analysis (ISODATA) clustering algorithm available in ERDAS/IMAGINE image processing software was used for the purpose. In the classified image the classes were visually assessed for their correctness and suitably labeled. Supervised classification was attempted for few islands based on the extensive ground survey done during 2001-2003. In certain cases visual interpretation was also performed with on-screen digitization. Digitized maps were edited, labeled and projected in the polyconic projection. The area was estimated in ha/sq km. The mangroves were classified based on density (Nayak and Bahuguna, 2001) into dense (> 40 % canopy cover) and sparse (< 40 % canopy cover). The coral reef habitat was classified based on the classification system (Table 3) evolved earlier (Bahuguna and Nayak, 1998). The classification system helps in classifying the coral reefs based on geomorphological zoning.

Tabel 3: Classification System

Level I	Level I	Level II	Level III
Coastal Wetlands	Fringing Reef	Reef flat	Sand on reef Mud on reef Detritus on reef
		Sand	Sandy beach
	Mangroves	Dense Sparse	
	Mudflat		
Others	Saline land		
Low Tide Line & High Tide Line			

Detailed ground surveys were conducted on selected location on the coasts of all the affected areas for observing the changes in geomorphological features.

3.4 Mangroves

Mangroves, trees whose tangled roots grow above ground in coastal swamps, are a unique habitat for migratory birds, monkeys, lizards and turtles, etc. They contribute directly to rural livelihoods by providing forest products including timber, poles, fuel wood and thatch for houses and indirectly by providing spawning grounds and nutrients for fish and shellfish. Mangroves, which lie in the inter-tidal region between sea and land, help to protect and stabilize coastlines and enrich coastal waters. The tsunami surge, which has gone through mangroves, ripped them completely, releasing silt, sediments, nutrients and pollutants. Table 4 gives the ecosystem mangrove area (in ha) affected by tsunami.

Mangroves of the Andaman and Nicobar Islands contribute about 18 per cent mangrove cover of India. The deeply indented coastline of the Andaman and Nicobar Islands results in innumerable creeks, bays and estuaries and facilitates the development of mangroves. In recent years it has been

Table 4: Ecosystem Area in the Andaman and Nicobar Islands affected by Tsunami

Ecosystem	**Area in ha**		**Change**	
	Pre-tsunami	Post-tsunami	In ha	Per cent
Mangroves	75078	59579	-15499	20.64
Coral	64364	13902	-50462	78.40
Reef- Reef Flat	5720	13264	+7544	131.89
Sand on reef Detritus	00	3768	+3768	-
Sandy Beach	5337	9326	+3989	74.74
Mudflats	12199	25804	13605	111.53
Saline Area	17585	59612	42027	239.00
Forest - Dense	NA	NA	14463	-
Open	NA	NA	637	-
Plantations	NA	NA	435	-

realized that mangroves may have a special role in supporting, stabilizing the coastal zone and protecting the lives and properties of the people living near the sea and onshore islands (Jackson and Winant, 1983; Jenkins and Skelly, 1987; Qureshi, 1990; Siddiqi and Khan, 1990; Mazda et al., 1997).

The total area under mangroves as mapped using satellite data is about 680 sq km in Andaman group of Islands and 70 sq km in Nicobar group of Islands (Nayak et al, 1992). The digital terrain elevation models clearly indicated that the low-lying areas have been affected badly barring the hilly terrains. In general, three zones of mangroves in this group of islands (Kannan et al. 2004), viz., proximal zone facing the sea, middle zone and the distal zone faced varied degree of damage.

During the large wave actions like tsunami, the near-shore (or proximal sea-facing) mangrove areas get damaged the most. Satellite data observations revealed that overall, the mangroves facing the sea have undergone most destruction. Extensive areas lie barren or have been totally eroded after tsunami (Fig. 1a). Similarly, typhoons in 1969/71 have devastated mangroves in Philippines (Walters, 2003).

Mangroves occupying the regions along the creeks in the entire Andaman group of islands have suffered comparatively less destruction. The distal zone mangroves have also suffered damage. Here the damage is due to prolonged inundation resulting in increase in salinity. The tidal flushing is during spring tides only and therefore the resident inundated water remained for a longer time here. In general, the inundation is higher at the heads of shallow, broad bays and inlets (Clague, 2001).

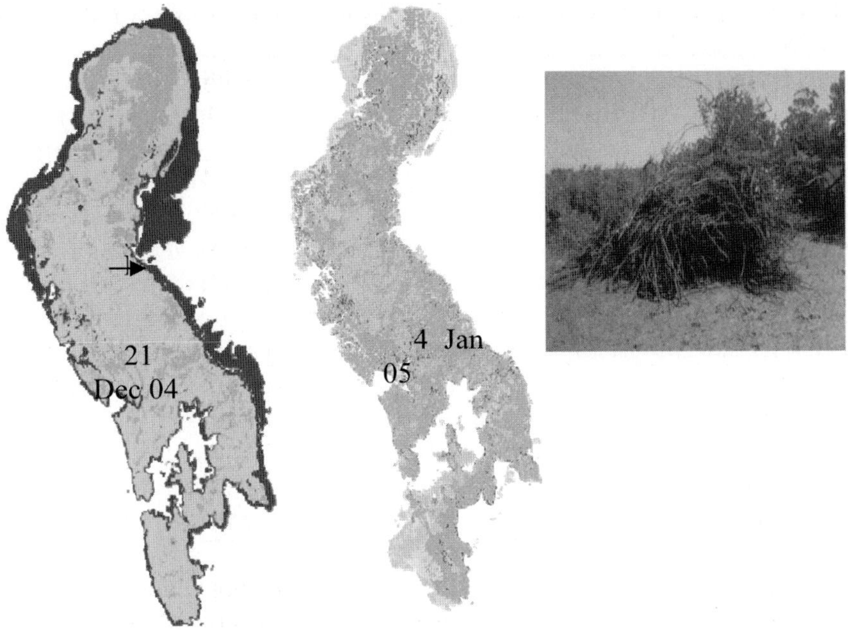

Fig. 1a. The sea-facing mangroves (M) present during pre-tsunami period (A) in the Comorta Island have been uprooted ( ) post-tsunami (B). 'C' is the ground photograph of the uprooted mangroves

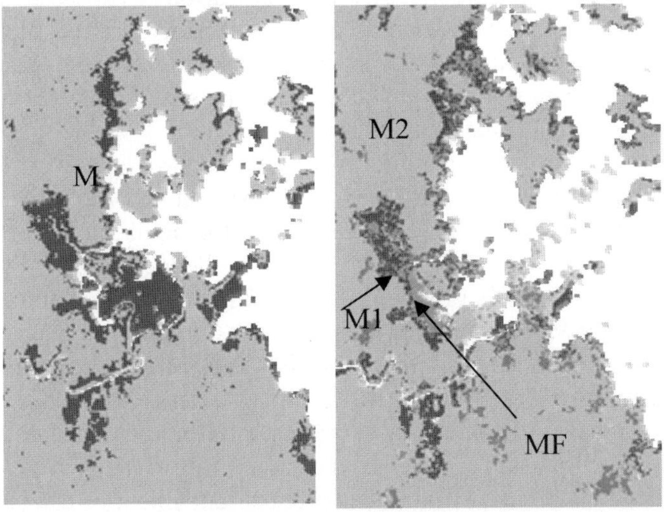

Fig. 1b. Damage to middle (M1) and distal zone (M2) mangroves of the Middle Andaman are seen in the post-tsunami map (D). The pre-tsunami map (C) shows dense mangroves (M). The barren and exposed mud is depicted as (MF).

The destruction to the mangroves of the Nicobar Islands was more on the western side than eastern side. This is due to presence of extensive reef formations along the eastern side of the Nicobar Islands which provided some protection. All these loss of mangroves would question the very existence of the endangered saltwater crocodiles that infested here. Mangroves of the Megapod Island have been totally washed out. Damage to this island especially to its vegetation would mean that the endemic Megapode birds would be critically affected, as these birds are not capable of flying.

There are 60 species of mangroves in the Andaman and Nicobar Islands making it the most diversely rich area in the Indian subcontinent (Nayak and Bahuguna, 2001). The dominant species are *Rhizophora apiculata, R. Mucronata, Brugeira gymnorrhiza, Heritiera littoralis, Nypa fruticans, Sonneratia caseolaris, Excoecaria agallocha* (most commonly distributed throughout the islands). *Avicennia sp* is present in the disturbed places on the Andaman Islands and is absent on the Nicobar Islands. The sea facing mangroves of these regions comprise pure stands of *Rhizophora* covering extensive areas (58% area). *Bruguiera, Xylocarpus* and *Ceriops* occupy the inward region (Nayak et al. 2003). After tsunami it was noted significantly that no damage has occurred to these mangroves. It was observed by Wolanski et al., (1992), that mangrove areas especially occupied by *Rhizophora* species suffered less damages. This may be due to the high density of *Rhizophora* trunks and roots in the bottom layer helping the pure stands of *Rhizophora* to withstand the impact of tsunami.

Mangroves are scattered all along the south-east coast of India and rich mangroves are present in Pichavaram area. In this region, no loss of forest or mangrove cover (*Avicennia marina, Rhizophora-Avicennia marina)* was observed. Pichavaram mangroves played a vital role in reducing the impact of tsunami wave actions on the nearby land. The seaward mangroves have been uprooted or tilted by strong tsunami waves. On the south-west coast, no major changes in mangrove cover were observed except in the one place where 8 ha (6 ha dense, 2 ha open) of mangrove cover have been lost due to tsunami waves.

Apart from the most visible damage of uprooting of the trees and decay due to inundation of water, mangrove ecosystems experience changes in their composition after an event like tsunami. Natural changes on mangrove coasts due to natural hazards include both seaward advance, with pioneer species spreading forward on accreting mudflats, and recession as the result of near shore and shoreline erosion (Talbot and Wilkinson, 2001). Where an advance is taking place, successional changes are usually in progress in the forest to the rear, with displacement by other swamp vegetation (e.g., *Nypa* palms) at the landward margin as accretion raises the substrate above high-tide level. The mangrove community thus migrates seawards on prograding coastlines. Further studies would be required on the mangroves of these islands to look for long-time impact of tsunami.

3.5 Coral reefs

Although corals occupy less than one quarter of one percent of the marine environment, coral reefs are home to more than a quarter of all known marine fish species. They provide food, livelihood and other essential services for millions of coastal dwellers, as critical fish habitat, popular destinations for eco-tourism, or protection to coastal communities from storms and hurricanes. Coral formations act as buffers during storm surges and tidal waves. When giant tsunami waves smashed onto shores, the coral in nearby shallow areas were destroyed, crushed and shrouded in debris.

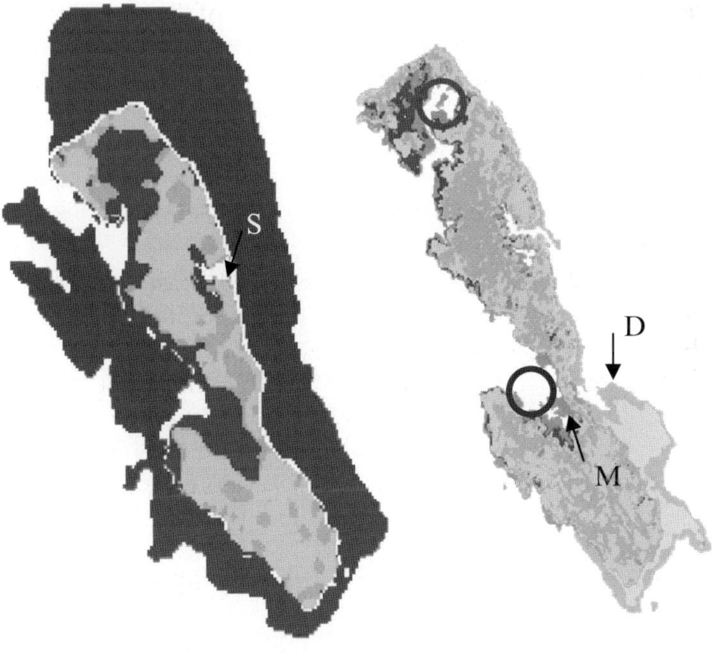

Fig. 2A. Status of coral reef of Trinkat Island during pre-tsunami is shown in (A) and post-tsunami period in (B). The reef area (R) and beach (SB) has been eroded in (B). The deposition of detritus (De) and sand (S) post-tsunami is also seen. The mangroves (M) that have been uprooted are shown in red circle in (B) and the mud laid barren and exposed is also seen (MF). Increase in saline area (SA) is also seen.

The massive backwash returned to sea carrying a deadly cargo of detritus material and deposited on coral reef. Siltation led to choking and death of the live coral reefs. The effects of coastal pollution, added to raw sewage released by the disaster, could further harm coral habitats and threaten already depleted fish stocks.

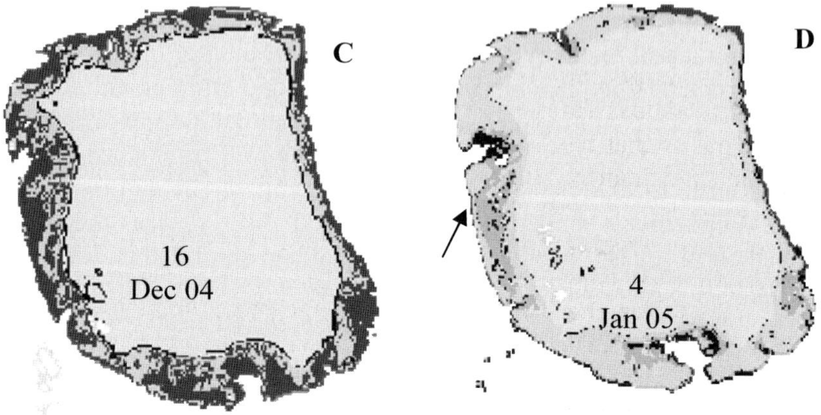

Fig. 2B. The damage to coral reef due to backwash is clearly seen on the Sentinel island reef. The pre-tsunami map (C) shows coral reef (R) surrounding the Sentinel Island. During Post-tsunami (D) the entire reef is covered with sand (S) and detritus (De)

The coral reef ecosystems are susceptible to damage from natural hazards such as tsunamis. They can be physically damaged and/or broken or covered with debris and sediment resulting in increased risk for marine invasions (IUCN 2005). Observations from satellite data also revealed two main kinds of damage to coral reefs, viz., i) total erosion or breaking up of reefs (Fig. 2A), and ii) deposition of sand, mud and detritus on reef (Fig. 2B). These two types of damage are in confirmation with the damage indicators of reef relevant to the tsunami, viz., i) wave damage – physical damage from the breaking waves, and ii) backwash damage – physical damage and physiological stress from materials and debris carried back over the reefs by the receding backwash off the land. The erosion of reefs can be explained due to wave damage and the degradation of reefs is due to backwash damage

The depositions prove to be as harmful as the erosion itself. Depositions led to blocking of the living corals resulting in their death. The detritus on reefs (that is the organic material) may also prove to be a favorable ground for the growth of filamentous algae and thick turf/fleshy macro-algae resulting in further degradation of these reefs.

The coral reefs of the Andaman and Nicobar group of islands are of fringing type with narrow reef flat in most of the places. The western side of the Islands has coral reef running intermittently for about 350 km. The Andaman and Nicobar Group of Islands sunk between 1 and 4 m and as a consequence large land areas have been submerged and are now to be found at 1 to 4 m depth. Many parts of the Andaman Islands have been uplifted causing large parts of reefs exposed (Kulkarni, 2005). As a consequence, a total loss (erosion) of 23000 ha of Andaman coral reefs and 17200 ha of Nicobar reefs as a result of tsunami was observed. About 6700 ha area in Andaman reefs and 6140 ha in Nicobar reefs is now covered with backwash.

The erosion to reefs is more pronounced in the Nicobar group of islands. The reefs of most islands mostly occupied the eastern side of the island. Coastal subsidence and high intensity waves along with nearness of these reefs to the epicenter may explain the massive erosion of these reefs. Maximum physical damage has been recorded in the reef edge region where even the massive *Porites* colonies have been upturned and damaged. Branching *Acropora* colonies have been broken. Earthquake has caused about 15 – 60 cm wide and 80 – 100 m long cracks in the massive live and dead coral to reef slope.

Sand deposition on the reef, detritus and mud on the reef are indicators of degrading/degraded reefs. All the reefs of the Nicobar group of islands have suffered pronounced deposition of sand, detritus and mud. Siltation was observed on the top of massive corals like *Porites* sp in reef flat areas, and a thin deposit of micro algae (about 30 %) was observed in recently dead coral due to tsunami. Sand has deposited on 1300 ha of reef and detritus covers 1200 ha of reef area in the Nicobar Islands after tsunami. The deposits could be as thick as 1 m.

It was observed that the reef slope and reef edge have lost their live coral cover more as compared to the reef flat region. The *Acropora* sp and branching corals have undergone most damage (40% damage) compared to massive *Porites*. Similarly damage to soft corals and fragile members of the reefs of Sri Lanka, Seychelles, Maldives, etc., due to the tsunami was also observed (IUCN, 2005, UNEP, 2005).

The impact of tsunami has been observed in the coastal region in the Gulf of Mannar (south-east coast). However, no significant impact was noticed on reefs, associated habitat and resources in the Gulf of Manner except minor transitional damages. Fine sand deposited (4-6 cm) in almost all cup corals has been observed in the mainland patchy reef after tsunami. Generally, 25-30% of cup corals in this area are now filled with fine sand (layers of 1-2 cm). Few table corals were tilted and branching corals were found to be broken. No impact on reef associated fishes, crustaceans and mollusks have been observed.

3.6 Coastal landforms /wetlands

The major coastal landforms/wetland features, which have been affected by tsunami, include beach, spits, sand dune, mud flat, tidal inlet, estuary, cliff, etc. A rapid assessment of damages to these systems has been made.

Beach ridges offers protection against flooding during freak high waves and wave set up due to long period and well grouped waves. Sandy beaches are the natural protection to the coasts and also the major area for the activities of the coastal communities. Most beaches/sandy areas were affected. They have been either washed away or sand was deposited on backshore areas and changed the beach configuration (Figs. 2A and 3). The details are given in Table 4.

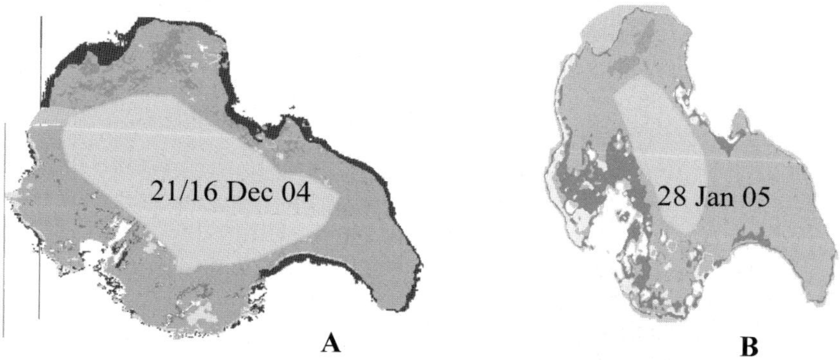

Fig. 3. Beach (Be) making up the shoreline of the bay south of Katchal island during pre-tsunami (A) period has been eroded post tsunami (B) image shows mud deposition (M)

Nesting beaches of the Nicobar group of islands have almost vanished as all these islands have gone down by 1-3 m. Every year, November to April is the nesting season for one of the species of the endangered leatherback turtles that visit the Great Nicobar Island and the Little Nicobar beaches, with the nesting peaks between December and January. As tsunami occurred during this period and the nesting beaches have been fully swept away from the shore, one generation of turtles would have been probably killed. Further nesting along these coasts may be nearly impossible as all these nesting beaches are now under sea.

On the mainland coast, the sediment from the beach was washed further landward in many places. Sediments got accumulated landward of backshore in almost all tsunami affected areas. Near Pulicat Lake, huge amount of shells have been deposited about 10 m from shoreline. The sand washed landward of backshore is almost lost to the beach. In many cases, the composition of beach sands has changed, e.g. white and coarser beach sand has changed to finer and black coloured sticky sand. The impact on beaches would be of transitional nature. These would reform within a short period of time. The deposition of heavy minerals, mainly ilmenite and garnet, were seen along parts of the south-east (Fig. 4) and the south-west coasts. They have been deposited up to 25 m from the shore. It was observed that generally coarser sand has been replaced by finer but heavier sand.

Fig. 4. Beach sand (black) washed landward of backshore at Alappad, Kola (Kerala)

Barrier islands/spits run parallel to the mainland coast but are separated from the mainland by lagoons and bays. They act as a sediment buffer for coastal protection and habitat for beach and inter-tidal organisms. The tsunami impact was severe at on the south-west coast. In many places barrier islands/beaches were breached opening new connections in the lagoon. The tsunami water overflowed across the barrier beaches in a number of locations along the south-east coast.

Breaching of permanent nature can induce shoreline changes. The breaching of barrier systems introduced high salinity to the backwater, which will have an impact on the brackish-water inhabitants (organisms). The ground water table also affected due to salinity intrusion. The impacts would have a long-term effect especially with respects to the contamination of the ground water table.

Sand dunes provide a physical barrier against storm waves, reducing the risk of flooding for nearby properties. These act as sand reserve for coastal stability. These are the habitats for many plants and animals and provide protection and shelter for supply and recharge of fresh water. Sand dunes are present along ports the south-east coast and along the south-west coast. Removal of sand and also deposition of sand on the backshore has been witnessed due to tsunami. The impact of tsunami on coastal sand dunes was severe along the south-east coast. Dunes were damaged along the south-west coast. Along this stretch, where man has tempered the dunes, and the damages were substantial. Damage to coastal properties was very less in places wherever dunes were present. Sand dunes have helped reduce the impact of the tsunami and this has helped protect the nearby settlements from being destroyed.

The loss of dunes will make the coastal communities more vulnerable to the hazards of coastal erosion and other episodic events. This will also reduce the fresh water recharge of groundwater table. Sand-binding plants play an important role in the formation, development and maintenance of dunes. Replanting of dune vegetation will help regeneration of dunes over a long period. Mining or tampering with the dune system should be totally banned. Development of artificial dunes will also stimulate dune regeneration.

Mudflats are formed by the deposition of finely inorganic material and organic debris in particulate form, which has been held in suspension in the sea or in estuaries. This environment is no less rich in invertebrate life than other littoral habitats. Moreover, mudflats possess special significance because they are the main feeding medium for many wading birds and wildfowl, which congregate on the coast in great numbers during migration periods and in the winter months. Other than being fishing and nursery grounds, these also act as sinks in controlling floods.

The mudflats present in all other sheltered regions have been significantly increased on all islands of the Andaman and Nicobar Group of Islands (Figs. 1B, 2A and 5). The increase in area is also due to destruction of mangroves. The effect on mudflats was substantial in many places along the south-east coast. In many locations, mudflats were covered with sand. However, along the south-west coast this was limited. The damages to mudflats have an impact on livelihood activities due to its effect on fishing and breeding/nursery grounds. The system will recoup within a few years.

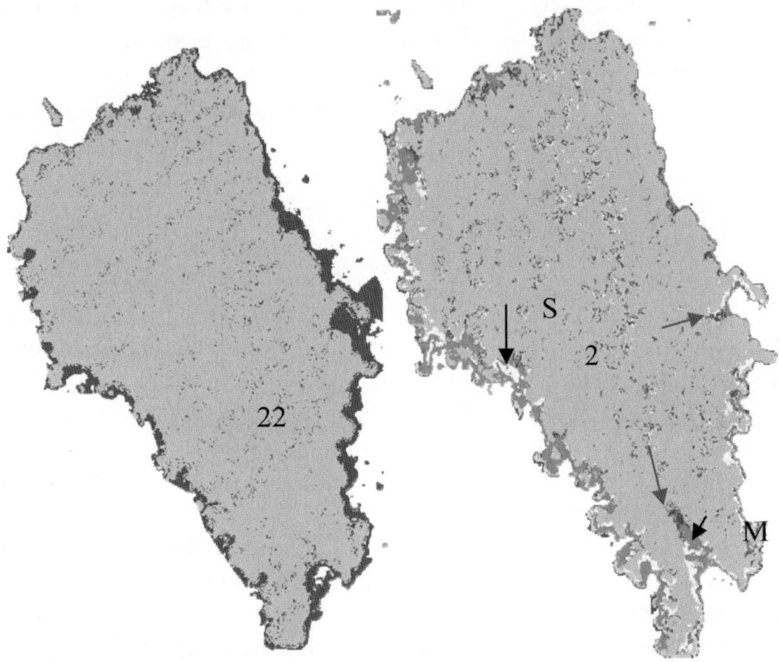

Fig. 5. The inundation of seawater landward (⟶) is seen in the Great Nicobar map of post-tsunami (B). Pre-tsunami (A) the area was forested land. The saline area (SA) as increased along with mud depositions (MF).

Tidal inlets play an important role in coastal ecosystems facilitating mixing of water, sediments, nutrients and organisms between terrestrial and marine environments. These are also the water routes across the coast for ships between harbours and the open sea. Tidal inlets such as lagoons and estuaries are highly productive ecosystems. Many species migrate into lagoon/estuarine system to feed, thereby taking advantage of the considerable production of organic matter and the lack of competing species. The major damages to tidal inlets were noticed along the south-east & south-west coast at few places. In many locations the tidal inlets, which were closed (usual characteristic of fair-weather season) got opened, e.g. Adyar estuary (Fig. 6). In some estuaries, the mouths have closed after tsunami. No permanent damages to livelihood activities are expected. The tidal inlets will regain its positions naturally. The shift in the location of tidal inlets can induce to shoreline changes. It requires further investigations to understand whether the breaches are permanent and to assess the possibility of enhanced shoreline changes.

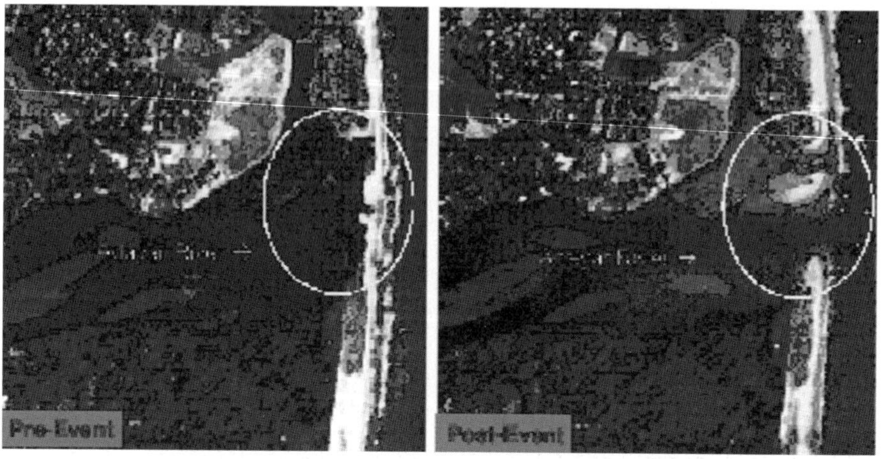

Fig. 6. The closed tidal inlet of Adyar estuary during pre-tsunami (A) period has opened up after tsunami (B).

The lagoon systems in all the affected areas on south-west coast were contaminated by highly saline water over wash or through new openings due to breaching of barrier beaches. There is also a possibility of a decrease in the depth due to the wash over of beach and near shore sediments. The survey indicated no marked changes in the Pulicat lake ecosystem. Only marginal sand deposition was observed. Further investigations are required for its long-term impacts on the livelihood activities.

Large tracts of land have become saline. It is especially noticeable in the South Andaman Island, Port Blair area and Great Nicobar (Fig. 5). The details are given in the Table 8.

3.7 Shoreline

The satellite images clearly show the shoreline changes. This study indicates that the shoreline along with the high tide line and low tide line has undergone noticeable changes in several areas (Fig. 7).

Fig. 7. Ground photograph of Aerial Bay, Diglipur in Andaman, shows the shift in shoreline after the episodic event.

The following types of shoreline change were noticed along the south-east coast.

- Regression of shoreline varying from 5–15 m
- Advancement of shoreline varying from 1-250 m, e. g. near Thiruvanmaiurkuppam and Reddiarpetta (~ 50 m), near Injabamkkam (~ 5 m),
- Development of new sand bars, e. g. in the Vellar river
- Erosion of sand bars

3.8 Forest

In the Andaman group of islands (Andaman District), no detectable change in forest cover was observed except in the Little Andaman Island where loss of 250 ha of forest cover has been assessed. In the Nicobar group of Islands significant damage in forest cover was detected. The results of the assessment are given in Table 4. An interesting observation made during the study was that changes in forest cover were more pronounced on western side of most of the islands than on the eastern side, which faced tsunami waves directly.

3.9 Conclusions and recommendations

The use of satellite data has facilitated a quick assessment of damage to the ecosystems of the Andaman and Nicobar Islands and mainland coast. AWiFS images are radiometrically better with 10 bits quantization level and LISS III data has 7 bits quantization level. It is mainly because of the better radiometry that the AWiFS data in spite of the coarser resolution could be used to assess the damage at the habitat level of the narrow coastal zone. The satellite data provided a record of damage assessment which was a vital input for estimating ~~of~~ the damage of type and magnitude, as well as recovery status, to plan for mitigation and restoration activities.

The reefs of the islands were in good condition prior to tsunami (Bahuguna and Nayak, 1998, Nayak, et al. 2003). These reefs have been most severely affected by the tsunami waves. The coral reefs have suffered erosion and also damage due to backwash. The erosion is most severe among the Nicobar reefs. Backwash has mainly resulted in deposition of sand, detritus (organic matter) and mud on the reefs. Reefs may recover from damage due to backwash when material gets washed out in due course of time. Long-standing deposition may cause irreversible damage to the reefs.

The sea facing mangroves have broken up or been uprooted. The sheltered mangroves have suffered less damage and have been degraded due to inundation. Long inundations of stagnant water cause death and decay of mangroves. The species harboring pneumatophores, knee roots, etc., (*Avicennia, Bruguiera)* may die a quicker death even in shorter durations of inundation as compared to aerial roots mangroves (*Rhizophora*). It has been noted that coasts harboring *Rhizophora* have not suffered less damage due to tsunami. The mangroves that have suffered loss due to land slumping are lost forever. Changes are expected in community structure as a result of tsunami.

The state of ecosystem affects directly the coastal communities as their livelihood is dependant on the resources provided by these ecosystems. The fishery has been adversely affected. The other major impact was limited availability of ground water as most groundwater resources turned saline. Many crop lands turned saline and became unproductive.

The other major impact was on some of endangered species such as saltwater crocodiles, leather-back turtles and Megapode bird as their habitat was lost. This damage is long-term and would take several years before habitats are restored. We need to develop multipurpose strategies incorporating the resilience of natural systems. It is quite possible that recovery can occur at a slow rate although composition and structure may change depending upon highly variable natural conditions.

3.10 Acknowledgements

The authors are grateful to Dr. P. Goel, Secretary, Ministry of Earth Sciences, Govt. of India, and Dr. R. R. Navalgund, Director, Space Applications Centre, Ahmedabad, for providing guidance and facility to carry out this work. We are also thankful to Central Agricultural Research Institute; Port Blair for providing us ground observations.

3.11 References:

Bahuguna Anjali and Nayak S. (1998), Coral reefs of the Indian coast, Scientific Note, SAC/RSA/RSAG/DOD-COS/SN/16/97 Space Applications Centre, Ahmedabad, p 56

Clague, J.J., (2001). 'Tsunamis,' In Brooks, G R (ed) A Synthesis of Geological Hazards in Canada. Geological Survey of Canada, Bulletin 548, p 27–42.

CRI/ISRS (2005). Tsunami Damage to Coral Reefs: Guidelines for Rapid Assessment and Monitoring. GCRMN/CORDIO/IUCN/Reefbase/ReefCheck/AIMS. January 2005. Seychelles. http://www.iucn.org/tsunami/temp/index.htm. Accessed: 12 July 2007

IUCN (2005) Recovery from the Indian Ocean tsunami - Guidance for ecosystem rehabilitation incorporating livelihoods concerns. Information Paper 1. IUCN. World Headquarters, Switzerland. http://www.iucn.org. Accessed: 24 Oct 2007

Jackson, G.A., Winant, C.D., (1983) Effect of a kelp forest on coastal currents. Continental Shelf Res., 2(1): p75-80.

Jenkins, S.A., Skelly, D.W. (1987). Hydrodynamics of artificial seaweed for shoreline protection., SIO, No. 87(16) , Scripps Inst. Oceano, p 66

Kannan, L., et al., (2004). 'Assessment of marine biological resources of the Great Nicobar Biosphere Reserve and its ecology'. Final Consolidated Report submitted to the Min. of Env. and Forests, Government of India, p 321 (Unpublished).

Kulkarni, S., (2005) Tsunami impact assessment of coral reefs in Andaman and Nicobar Islands. Reef Watch Marine Conservation Mumbai, India. Interim Report, CORDIO (Unpublished).

Mazda, Y., Magi, M., Kogo, M., Hong, P.N., (1997) Mangroves as a coastal protection from waves in the Tong King Delta, Vietnam. Mangrove Salt Marshes, 1(2): p127-135.

Nayak, S., Bahuguna Anjali, Shaikh, M G, Chauhan, H B, Rao, R S, Arya, A S, (1992) Coastal *Environment*, Scientific Note, RSAM/SAC/COM/SN/11/92, Space Applications Centre, Ahmedabad, p 114.

Nayak S. and Bahuguna Anjali, (2001), 'Application of RS data to monitor Mangroves and other coastal vegetation of India', Indian J. Marine Sciences, 30(4): p195-213.

Nayak, S., Bahuguna Anjali, Shah, D.G., et al (2003) Community zonation of selected Mangrove habitats of India using satellite data. *Scientific Note*. Space Applications Centre, Ahmedabad, p 108

Qureshi, T.M., (1990) Experimental plantation of rehabilitation of mangrove forests in Pakistan. Mangrove Ecosystems Occasional Papers, vol. 4, p. 37, UNESCO, Paris.

Siddigi, N.A., Khan, M.A.S (1990) Two papers on mangrove plantations in Bangladesh. Mangrove Ecosystems Occasional Papers, vol. 8, p19 UNESCO, Paris.

Talbot, F and Wilkinson, C (2001) Coral reefs, mangroves and sea grasses: A source-book for managers, Joint ICRI/AIMS/GRCMN/GBRRF/IUCN/CORDIO/WWF publication.

UNEP (2005) Executive summary report on rapid assessment of tsunami: Effect of the 2004 Indian Ocean earthquake on India. http://en.wikipedia.org/wiki . Accessed: 24 Oct 2007

Walters, B.B (2003) People and mangroves in the Philippines: Fifty years of coastal environmental change, Environmental Conservation 30 (2): 293–303.

Wolanski, E., Mazda, Y., Ridd, P., (1992) Mangrove hydrodynamics. In: Robertson, A.I., Alongi, D.M. (ed), Tropical mangrove ecosystems. American Geophysical Union, Washington, p. 43-62.

4. Increasing the Use of Geospatial Technologies for Emergency Response and Disaster Rehabilitation in Developing Countries

David Stevens

Abstract

This paper discusses the notable increase in the availability of space-based data and information to support emergency response, and to a lesser extent disaster rehabilitation and preparedness, and the need to ensure that developing countries are able to access and use space-based technologies (earth observation and meteorological satellites, communication satellites and global navigation satellite systems) for risk reduction and disaster management.

Major initiatives that are helping developing countries are outlined such as the work carried out by the United Nations Office for Outer Space Affairs, the creation of the International Charter Space and Major Disasters, and the work carried out by the Committee on the Peaceful Uses of Outer Space (COPUOS). This last effort led to the United Nations General Assembly establishing, on 14 December 2006, the United Nations Platform for Space-based Information for Disaster Management and Emergency Response (UN-SPIDER) as a program of the United Nations Office for Outer Space Affairs.

The paper concludes by stating that even though there has been an increase in the number of international initiatives and the establishment of a global coordinating platform there is still a need for national institutions that have competence in the use of geospatial technologies to step forward and take a leading role in helping end-users incorporate such solutions in their daily operations.

4.1 Introduction

The notable increase in the availability of space-based data and information to support emergency response and to a lesser extent disaster rehabilitation and preparedness in developing countries is relatively recent, having occurred after the Third United Nations Conference on the Exploration and Peaceful Uses of Outer Space (UNISPACE III), which took place in 1999. This United Nations Conference highlighted the need to help developing countries have access and use space-based technologies (earth observation and meteorological satellites, communication satellites and global navigation satellite systems) for risk reduction and disaster management. The recognition of this need led to a number of international initiatives that together have been successfully contributing to helping developing countries access and use space-based information for disaster management and emergency response, such as, the work carried out by the United Nations Office for Outer Space Affairs, the creation of the International Charter Space and Major Disasters, and the work carried out by the Committee on the Peaceful Uses of Outer Space (COPUOS), which led to the United Nations General Assembly establishing, on 14 December 2006, the United Nations Platform for Space-based Information for Disaster Management and Emergency Response (UN-SPIDER) as a program of the United Nations Office for Outer Space Affairs. Even with the increase of the number of international initiatives and the establishment of a global coordinating platform there is still a need for national institutions that have competence in the use of geospatial technologies to step forward and take a leading role in helping end-users incorporate such solutions in their daily operations.

4.2 Geospatial technology solutions

Earth observation satellites have demonstrated their utility in providing data and information for a wide range of applications in disaster management (Tralli et al. 2005). Pre-disaster uses include risk analysis and mapping; disaster warning, such as cyclone tracking, drought monitoring, the extent of damage due to volcanic eruptions, oil spills, forest fires and the spread of desertification; and disaster assessment, including flood monitoring and assessment (Zhang et al. 2002), estimation of crop and forestry damages and monitoring of land use/change in the aftermath of disasters. Remotely sensed data also provide a historical database from which hazard maps can be compiled, indicating which areas are potentially vulnerable. Information from satellites is often combined with other relevant data in geographical information systems (GIS) in order to carry out risk analysis and assessment. GIS analytical tools can be used to model various hazard and risk scenarios for planning the future development of an area.

Meteorological satellites can monitor weather patterns, detect and track storms and monitor frosts and floods. Derived products are produced routinely several times per day, many of them focused on particular hazard events. Tracking sequences of tropical cyclone images from geostationary satellites as well as storm intensities and atmospheric winds derived from these images provide vital information for forecasting landfall, thus contributing to saving lives.

Global navigation satellite systems (GNSS) provide accurate position, velocity and time information that is readily accessible at ground level to anyone with a receiver, enabling the collection of data to support risk reduction and emergency response activities. Communication satellites are geo-stationary satellites that enable the setting up of emergency communication channels and are being used increasingly by all those responding to an emergency. Restoring communication in disaster stricken areas is usually the first priority when responding to an emergency. Additionally, there is a need to receive information from and send information to the various emergency response teams working in the field, including large data files such as maps and satellite images.

4.2.1 World conference on disaster reduction

The importance of space technology to support disaster preparedness was recognized at the World Conference on Disaster Reduction, held in Kobe, Japan, in January 2005. Representatives from 168 countries recognized the contribution of space technology to disaster reduction and emphasized the need to incorporate space-based services routinely to support risk reduction. The Conference was the largest gathering ever of the disaster community, totaling 4,000 participants in the plenary activities and the thematic sessions and around 40,000 people in the public segment. A list of commitments is set out in the main document produced, the 'Hyogo Framework of Action 2005-2015: Building the Resilience of Nations and Communities to Disasters', which was endorsed by the Governments at the Conference and adopted by the General Assembly in 2005 (United Nations, 2005a).

The Secretariat of the International Strategy for Disaster Reduction (ISDR) is working on the implementation of the Hyogo Framework, which is seen as the international roadmap for disaster risk reduction over the next 10 years. The Hyogo Framework recognizes the need to 'promote the use, application and affordability of recent information, communications and space-based technologies and related services, as well as earth observations, to support disaster risk reduction, particularly for training and for the sharing and dissemination of information among different categories of users'.

4.2.2 UNISPACE III

Under the theme 'Space benefits for humanity in the twenty-first century', UNISPACE III was held in Vienna from 19 to 30 July 1999. The most important result of this United Nations Conference was the adoption of the Vienna Declaration on Space and Human Development (United Nations, 2000) in which 33 specific actions were recommended that should be carried out to enable space technologies to contribute to the global challenges of the new millennium. One of the recommendations put forward was the need 'to implement an integrated, global system, especially through international cooperation, to manage natural disaster mitigation, relief and prevention efforts, especially of an international nature, through Earth observation, communications and other space-based services, making maximum use of existing capabilities and filling gaps in worldwide satellite coverage'.

The recognition of this need has led to a number of international initiatives that together have been successfully contributing to helping developing countries access and use space-based technology solutions for disaster management and emergency response, such as, the work carried out by the United Nations Office for Outer Space Affairs, the creation of the International Charter Space and Major Disasters, and the work carried out by the Committee on the Peaceful Uses of Outer Space (COPUOS).

4.3 Building a strategy

Building upon the priorities established as a result of UNISPACE III, the United Nations Office for Outer Space Affairs (UNOOSA) organized a series of regional workshops on the use of space technology for disaster management between 2000 and 2004, bringing the results of these regional workshops to a final international workshop, which was held in Munich, Germany, in October 2004. At that meeting, a total of 170 participants from 51 countries defined a global strategy that would contribute to helping developing countries gain access to and learn how to use space technology for disaster management, a strategy put forward as 'The Munich Vision: a Global Strategy for Improved Risk Reduction and Disaster Management Using Space Technologies' (United Nations, 2005b).

Participants at the Munich workshop recognized that space-based technologies such as Earth observation satellites, communication satellites, meteorological satellites and GNSS played an important role in risk reduction and disaster management, and put forward a number of recognitions and recommendations in the areas of capacity development and knowledge-building; data access, data availability and information extraction; enhancing awareness; and national, regional and global coordination.

With regard to capacity development and knowledge building participants recognized that it was the responsibility of the space technology community to reach out to understand the specific needs of the user community and develop end-to-end solutions that meet their requirements. Furthermore, there was a need for continuous education and training in space science and technology (at the technical, institutional and decision-making levels) and the development and consolidation of national and regional expertise.

The discussion on data access and availability and information extraction concluded that there is limited or no mechanisms in place to make data rapidly available at all decision levels during disaster response and when data are available it is not always in a 'user friendly' format. Participants recommended the development of standards for information extraction from remotely sensed data and disaster mapping procedures, which in turn would foster better understanding and acceptance of space-based information by civil protection and disaster relief communities.

Finally, the discussion on the need to enhance awareness led to the recognition of the need to create awareness among national and international stakeholders that the incorporation of space-based solutions reduces risk and vulnerability and are cost effective. Furthermore, lessons learned from the application of space-based technologies for the mitigation of hazards should be disseminated to the public, beginning with school children and including the scientific community and the media. Additionally, institutions within each country that use space technology should take on the responsibility to periodically carry out activities that contribute to raising awareness, such as promoting World Space Week (held 4-10 October annually).

Participants agreed that at the national level, institutions within a country should be responsible for coming together to define actions to be carried out collectively, with the appropriate space technology institutions responsible for taking the lead. Particular attention should be given to the local communities, involving local leaderships and grassroots organizations.

At the regional level, interested international, regional and national institutions should come together to advance the actions that are relevant to the region as a whole. At the global level, participants recognized the importance and urgent need of the coordination entity, which should be seen as a one-stop shop for knowledge and information sharing (best practices) and also as a platform for fostering alliances.

In fact, the establishment of such a global coordinating mechanism was one of the leading discussion topics taking place at COPUOS and its Scientific and Technical Subcommittee.

4.4 Establishing the UN-SPIDER Program

4.4.1 COPUOS

Following UNISPACE III, COPUOS agreed to establish action teams composed of interested Member States in order to implement the recommendations agreed to in 'The Vienna Declaration'. One of the action teams established focused on disaster management. This Action Team on Disaster Management was co-chaired by Canada, China and France, with the United Nations Office for Outer Space Affairs providing substantive assistance and secretariat services. The Action Team brought together 41 Member States and 13 intergovernmental and non-governmental organizations.

In its final report to COPUOS this Action Team made a number of recommendations, including the recommendation that an international space coordination body for disaster management should be established. Its recommendations, including the recommendation to establish such a coordinating body, were included in the report of the Committee on the Peaceful Uses of Outer Space on the five-year review of the implementation of the recommendations of UNISPACE III submitted to the United Nations General Assembly at its fifty-ninth session (United Nations, 2004).

Subsequently, at the request of the United Nations General Assembly, an ad hoc expert group was established in February 2005 bringing together experts from 26 Member States and from five specialized agencies of the United Nations and non-governmental organizations to continue the work of the Action Team by carrying out a study on the possibility of creating an international entity to provide for coordination and the means of realistically optimizing the effectiveness of space-based services for use in disaster management. This ad hoc expert group finalized its report early 2006 and presented it to COPUOS for its review during its forty-ninth session in June 2006.

4.4.2 The UN-SPIDER Program

The ad hoc expert group identified a number of important space-related initiatives, either ongoing or planned, that could support different phases of disaster management (i.e. risk reduction, prevention, mitigation, early warning, relief and rehabilitation), such as the Global Earth Observation System of Systems – GEOSS, the International Charter 'Space and Major Disasters', the Integrated Global Observing Strategy Partnership - IGOS-P, the Global Monitoring for Environment and Security - GMES and Sentinel Asia. A number of additional relevant initiatives such as RESPOND of the Global Monitoring for Environment and Security, Map Action and Global Map Aid were also identified.

The ad hoc expert group concluded, however, that gaps existed in the awareness of the disaster management community of the availability of such resources, in knowledge of how to access them and in the capacity to use them, also finding that these activities are driven by different mandates, often with a focus on specific disaster phases or types of crisis. There is no single, global coordination mechanism to implement an integrated disaster monitoring system that makes maximum use of available space technologies and services for the full disaster management cycle as called for in 'The Space Millennium: Vienna Declaration on Space and Human Development'.

The experts concluded that because such a coordination mechanism is needed an entity should be created with the following mission statement: 'Ensure that all countries have access to and use all types of space-based information to support the full disaster management cycle'. Furthermore, the ad hoc expert group recommended that such an entity be implemented as a program of the United Nations Office for Outer Space Affairs. The proposed name of the new program is: 'United Nations Platform for Space-based Information for Disaster Management and Emergency Response (UN-SPIDER).'

Subsequently the United Nations General Assembly agreed to establish UN-SPIDER, on 14 December 2006, as a program within the United Nations to provide universal access to all types of space-based information and services relevant to disaster management by: being a gateway to space information for disaster management support; serving as a bridge to connect the disaster management and space communities; and, being a facilitator of capacity-building and institutional strengthening.

The UN-SPIDER program is being implemented by the Vienna-based United Nations Office for Outer Space Affairs (UNOOSA) and additionally the new program will also have offices in Beijing (China) and Bonn (Germany) as well as a Liaison Office in Geneva (Switzerland).

Furthermore UN-SPIDER will work closely with end-users, particularly in developing countries, through the consolidation of a network of regional support offices (hence the appropriateness of the chosen acronym), building upon the commitments being provided by many countries and thus ensuring that regional and national centers have a strong role in their respective region.

4.5 Building upon existing initiatives

In 1999 when UNISPACE III delegates first highlighted the need for a space-based disaster management support system, the first thought was that there was the need for a dedicated global and coordinated system to be designed, built, launched and made operational with the support of interested countries. The fact is since then several initiatives have already started providing support to specific phases of the disaster management cycle (mostly though the response phase) and for specific geographic regions and countries. Additionally, several initiatives are working on coordinating the existing and planned space-based assets.

As the experts of the ad hoc expert group concluded, and as had already been put forward by the participants of the Munich workshop, what is missing now is a coordinating entity that brings together the disaster management and the geospatial communities. UN-SPIDER will be this coordinating entity, helping developing countries take advantage of existing initiatives such as the ones described below.

4.5.1 The International Charter Space and Major Disasters

An example of an initiative specific to a particular phase of the disaster cycle is the Charter on Cooperation to Achieve the Coordinated Use of Space Facilities in the Event of Natural or Technological Disasters (the International Charter 'Space and Major Disasters'). At UNISPACE III, a proposal to create a charter that would provide a unified system of space data acquisition and delivery to those affected by natural or man-made disasters through authorized users was put forward by the European Space Agency (ESA) and the Centre national d'études spatiales (CNES) of France, who signed the Charter in June 2000, with the Canadian Space Agency signing the Charter shortly after. Currently the Charter has expanded to ten members.

By the end of 2007, the Charter had been activated over 150 times in response to landslides, oil spills, floods, volcanic eruptions, forest and bush fires, earthquakes, storms and hurricanes, the Indian Ocean tsunami, the massive earthquake in India and Pakistan and even for hurricane Katrina in the United States. Specifically during the Indian Ocean tsunami, the Charter was activated a total of three times: firstly by India (through the Indian

Space Research Organization - ISRO) for its own territory, by the French Civil Protection Agency for Sri Lanka and finally by the United Nations Office for Outer Space Affairs for Indonesia and Thailand. This unified system played a major role in the aftermath of the Indian Ocean tsunami, having made available over 200 satellite images of the impacted areas to the emergency response efforts

The activation of the Charter by the United Nations, a total of 43 times by the end of 2007, has been made possible because the United Nations Office for Outer Space Affairs was accepted as a Cooperating Body to the Charter in March 2003, a mechanism through which the United Nations system can request imagery from Charter members by faxing in requests for imagery to support emergency disaster response situations. The Charter has successfully demonstrated the importance and the benefit of receiving timely satellite-based information to support emergency response. The need now is to expand this opportunity to the vast majority of countries that still cannot access the Charter due to the current activation rules of the Charter members.

4.5.2 GEOSS

One recent effort that is seeking comprehensive coordination and political commitment to guarantee the wide incorporation and use of space-based technology products and solutions is the 10-year implementation plan for a Global Earth Observation System of Systems (GEOSS) put forward by the intergovernmental Group on Earth Observations (GEO). The plan summarizes the steps that should be taken over the next decade by a growing community of nations and intergovernmental, international and regional organizations to put GEOSS in place. GEOSS will contribute to the establishment of the capacity to continuously monitor the state of the Earth, increased understanding of dynamic Earth processes, enhanced prediction of the Earth system and further implementation of international environmental treaty obligations. The plan includes consideration of the need for coordination of a wide range of space-, air-, land- and ocean-based environmental monitoring platforms, resources and networks—at present often operating independently.

The 10-year implementation plan recognizes disasters as one of the main areas that would benefit from such a coordination effort, contributing to reducing loss of life and property from natural and human-induced disasters. The implementation of GEOSS will bring a more timely dissemination of information through better coordinated systems for hazard monitoring, prediction, risk assessment, early warning, mitigation and response at the local, national, regional and global levels.

4.5.3 GMES

Similarly, the Global Monitoring for Environment and Security (GMES) is a joint initiative of the European Commission and ESA, designed to establish a European capacity for the provision and use of operational information for global monitoring of the environment and security. The overall aim of GMES is to support Europe's goals regarding sustainable development and global governance, in support of environmental and security policies, by facilitating and fostering the timely provision of high-quality data, information and knowledge, by means of three components: a partnership of key European actors, a European shared information system and a mechanism for permanent dialogue. By 2008 the foundations and the structuring elements of the European capacity for GMES should be in place and operating.

4.5.4 IGOS-P

The Integrated Global Observing Strategy Partnership (IGOS-P) is an international partnership, established in June 1998, which brings together a number of international organizations concerned with the observational component of global environmental issues, from a research point of view as well as from an operational perspective. The IGOS GeoHazards Theme is a combined initiative of three IGOS members, UNESCO, the Committee on Earth Observation Satellites (CEOS) and the International Council for Science. It intends to respond to the scientific and operational information needs related to the prediction and monitoring of geophysical hazards, namely earthquakes, volcanoes and ground instabilities. The main goal of the initiative is to investigate and develop an integrated observational strategy that would greatly enhance the operational and research capabilities of end-user agencies involved in mitigation of geohazards at the local, national and regional levels.

4.5.5 Sentinel Asia

Sentinel Asia is a regional cooperation project, with strong commitment from the Japanese Aerospace Exploration Agency (JAXA) and the participation of many international organizations, including the United Nations

Office for Outer Space Affairs, which aims at supporting the Asia-Pacific region to have access and use space-based products to support disaster management activities.

Through Sentinel Asia participating institutions from within the region are able to access data from the Japanese Advanced Land Observing Satellite (DAICHI) to support emergencies as well as access other widely available satellite data for fire and flood monitoring. The project also provides capacity building to these institutions so they are able to use such data and information for disaster management.

4.5.6 Constellation of disaster monitoring satellites

South Africa is planning to launch its first low earth orbiting satellite for disaster monitoring and other purposes. The satellite, named Sumbandila, which in local Tshivenda language means 'lead the way', will have a total cost of 3.7 million U.S. dollars, will weigh only 80 kg and be about 1.8 meters long (Africast, 2006). This reflects the trend towards a constellation of smaller satellites, which will contribute to the densification of satellites in space increasing the number of watchful eyes, and diminishing significantly the revisit time over any spot on the planet.

Once such constellation of 5 satellites already provides a daily revisit to any part of the globe: the Disaster Monitoring Constellation (DMC). The satellites, belonging to five countries (Algeria, China, Nigeria, Turkey and the United Kingdom), have already demonstrated the importance of daily revisit during several recent major disasters. Similarly there are other initiatives such as the Italian COSMO-SkyMed (Constellation of Small Satellites for Mediterranean basin Observation) and the disaster monitoring constellations to be launched by China and also the Russian Federation.

There is, and will continue to be, an increasing amount of space-based information being made available to the disaster management community. Initiatives such as GEOSS, GMES, IGOS-P and Sentinel Asia are helping the space community organize themselves to be able to take on a stronger role in solving the problems of the planet, including the increasing number of disasters. Initiatives such as the International Charter Space and Major Disasters and the existing Disaster Monitoring Constellation are coordinating existing space-based sensors so as to provide on demand access to space-based information for emergency response. But there is still the need to help the end user transform such information into usable products and incorporate these products into their activities.

Indeed, the Charter has made Earth observation data available to emergency relief teams, but an analysis of the impact of the data provided has shown the need for a full range of end-to-end services, including data processing and interpretation and not only satellite imagery. Several initiatives are filling that gap, such as RESPOND and the facility for the Center for Satellite Based Crisis Information (ZKI) of the German Aerospace Center (DLR).

4.5.7 Providing value-added services

RESPOND is an alliance of European and international organizations working with the humanitarian community to improve access to maps, satellite imagery and geographical information. RESPOND works during all phases of the disaster cycle where geographical information helps deploy humanitarian and development aid, paving the way for a set of sustainable services. It has been set up to identify the space-based information that is regularly used by humanitarian agencies when anticipating or responding to disasters. In addition to base mapping and satellite-derived information, RESPOND is also committed to supporting training, providing support services and infrastructure, forecasting and alert services, thus covering a large part of the disaster management cycle. The services are intended to respond to slow-onset disasters such as famine and desertification, as well as to sudden emergencies such as tsunamis, earthquakes and floods.

ZKI is a service of the German Remote Sensing Data Center (DFD) of DLR. It is responsible for providing rapid acquisition, processing and analysis of satellite data and the provision of satellite-based information products on natural and environmental disasters, for humanitarian relief activities, as well as in the context of civil security. The analyses are tailored to meet the specific requirements of national and international political bodies as well as those of humanitarian relief organizations.

4.6 Supporting end users

The Tsunami that swept through Southeast Asia late 2004 clearly demonstrated that space-based data and information are increasingly being made available to support emergency response. But it also demonstrated that developing countries still do not have wide access and do not know how to use such data and information, not only during the response phase, but also during the more important preparedness phase of the disaster cycle, which invariably results in unnecessary loss of life and property when disaster strikes.

4.6.1 Taking the leading role

The UN-SPIDER Program is playing a major role in helping developing countries access and use all types of space-based information to support the full disaster management cycle. It is doing this by being a gateway to space-based information for disaster management support, by being a bridge that connects the disaster management and space communities and by being a facilitator of capacity building and institutional strengthening. But ultimately the responsibility for ensuring that UN-SPIDER is successful at the local level will fall upon the national institutions that step forward and take the lead.

The Program is working closely with National Focal Points (NFPs) that are nominated by their respective Governments, in the implementation of risk reduction and emergency response activities and projects identified in conjunction with these NFPs. But ultimately the responsibility to ensure that each country takes advantage of existing space-based technology solutions will fall upon the existing relevant national institutions.

Even with the increase of the number of international initiatives and the establishment of a global coordinating platform there is still a need for national institutions that have competence in the use of geospatial technologies to step forward and take a leading role in helping end-users incorporate such solutions in their daily operations.

These national institutions have to take a leading role and ensure:

- Systematic compilation of relevant geospatial data and information or information on where such data and information can be obtained.
- Compilation of risk information at the national level by disaster theme and the development of national vulnerability assessments with the support of geospatial technologies.
- Awareness raising and outreach activities contributing to helping end-users understand the importance of incorporating space-based technology solutions in risk reduction and disaster management.
- Support to Capacity Building: the use of new technologies is only possible if experts are provided with the needed training. Capacity building and strengthening of institutional arrangements at all levels is key to increasing the ability of organisations and individuals to effectively use space-based services for disaster preparedness, response and recovery.
- Consolidation of a knowledge-base – identify and bring together knowledge, practical know-how, expertise and best practices, focusing on capturing and making such knowledge available to all end-users. This includes the refinement of user requirements and definition of best practices.

- Establishment of Communities of Practice as mechanisms to bring together the end-user and the providers of space-based solutions.
- Coordination at a national level among relevant national institutions, scientific institutions, organisations implementing and/or providing space-based solutions, humanitarian, environmental and civil protection actors and the space community.
- Establishment of a National Disaster Management Planning and Policy that builds upon the Hyogo Framework of Action and that recognises the role of space-based technologies.

4.6.2 Final thoughts

Geospatial information is being made available at an increasing rate to support end users of the disaster management and emergency response community. New initiatives are being implemented that target specific hazards or steps of the disaster management cycle. In 2007 a global coordinating entity, the UN-SPIDER Program, was established contributing to bringing the end user closer to the space-based solutions being made available. But ultimately the day-to-day help these end users need to be able to understand such solutions and incorporate them in their disaster management activities will come from institutions that are in the best strategic position to provide such support.

References

Africast (2006) SA to launch satellite in December, 1 August 2006, http://news.africast.com/africastv/article.php?news ID=59456 (accessed 02 August 2006).

Tralli DM, RG Blom, et al. (2005) Satellite Remote Sensing of Earthquake, Volcano, Flood, Landslide and Coastal Inundation Hazards. ISPRS Journal of Photogrammetry & Remote Sensing 59: 185-198.

United Nations (2000) UNISPACE III: A Strategy for International Cooperation to Promote Space Benefits for Humanity, August 2000, (United Nations publication, Sales number: 00.I.13.

United Nations (2004) Review of the Implementation of the Recommendations of the Third United Nations Conference on the Exploration and Peaceful Uses of Outer Space, A/59/174, 23 July 2004, http://www.unoosa.org/pdf/reports/unispace/A_59_174E.pdf (accessed 02 August 2006).

United Nations (2005a) Report of the World Conference on Disaster Reduction, A/CONF.206/6, 16 March 2005, http://www.unisdr.org/eng/hfa/docs/Final-report-conference.pdf (accessed 02 August 2006).

United Nations (2005b), United Nations International Workshop on the Use of Space Technology for Disaster Management, Munich, Munich, 18-22 October 2004, A/AC.105/837, 23 May 2005, http://www.unoosa.org/pdf/reports/ac105/AC 105_837E.pdf (accessed 02 August 2006).

Zhang J, C Zhou, et al. (2002) Flood Disaster Monitoring and Evaluation in China. Environmental Hazards 4: 33-43.

Part 2: Remote Sensing Technology for Disaster Monitoring

5. Adopting Multisensor Remote Sensing Datasets and Coupled Models for Disaster Management

Gilbert L. Rochon, Dev Niyogi, Alok Chaturvedi, Rajarathinam Arangarasan, Krishna Madhavan, Larry Biehl, Joseph Quansah and Souleymane Fall

Abstract

An application and a process involving integration of dynamic models for a data-rich environment, incorporating a multi sensor dataset is discussed. The potential utility of such data fusion for different phases of disaster management: vulnerability assessment, early warning systems, disaster mitigation, response, damage assessment and recovery are delineated. Case studies are drawn from disaster scenarios for flooding, drought management, and heavy rains in India. Applicability of the technology and processes, with potentially different sources of data, is described. Solutions to several technological challenges to handle large data sets using distributed cluster technology and data visualization, using high-resolution large display systems, are presented. Taking an example of the July 26, 2005 heavy rain events in Mumbai, India, which caused flooding, and resulted in over 400 deaths and nearly a billion US$ economic losses, the ability of multiple models to study the predictability, variability and use of model – satellite data fusion for severe weather and disaster mitigation, as well as response needs is discussed. A case for multisensory satellite datasets and the use of upcoming technologies, including handheld computers and cell phones in facilitating early warning, evacuation and emergency intervention is addressed. A case is made for a technological and educational infrastructure development that can benefit from remote sensing centric models with different complexity and a community cyberinfrastructure for multi-data access for disaster management.

5.1 Introduction

Several of the so-called 'natural' or 'biogenic' disasters can have human-induced (i.e. anthropogenic) components. These include an array of disasters associated with or exacerbated by climate change. Other disasters are also the result of human impacts, such as pollution from industry, passive run-off of agricultural pesticides and fertilizers, urban effluent, hazardous chemical and oil spills, deliberately or inadvertently set forest fires, and acts of terror. The literature is replete with case studies of remote sensing applications for these tragic human-caused events. One example of human suffering caused by other humans is the prevalence of landmines and unexploded ordinance. Filippidis (1999) and Habib (2001) both address the applicability of remote sensing and algorithm development for mine detection.

The global population is becoming increasingly susceptible to natural disasters. It would be disingenuous to argue that remote sensing alone offers salvation from such events; however it can legitimately be argued that remote sensing has emerged as one set of tools that can be used to mitigate the impact of natural disasters. Examples include facilitating disaster planning, providing early warning, enabling vulnerability assessment by expediting population evacuation and appropriate emergency response, and improving damage assessment capability, post-disaster humanitarian assistance and subsequent reconstruction of infrastructure.

This chapter provides an overview of different remote sensing techniques and data fusion applications with multiscale models. The following section overviews the broader environmental sustainability projects and vulnerability assessment applications adopting remote sensing. Section 3 provides an overview of the emergency warning systems (EWS) and the remote sensing products typically available for the model data fusion applications. Section 4 provides a case study of a model data fusion application for a record breaking heavy rain event over the Indian monsoon region. The advent of new technologies and the challenges of cyberinfrastructure for multisensory data access are outlined in Section 4. The educational and training needs are outlined in Section 5. The chapter builds on a summary workshop hosted in India. Therefore many of the examples discussed are targeted towards the Asian region. However, the concepts and data fusion techniques or applications are generally global in nature.

5.2 Remote Sensing in Disaster Management and Sustainability/ Vulnerability studies

The devastating impacts of the December 2004 Tsunami and earthquakes in Indonesia and in Pakistan and 2005 Hurricanes Katrina and Rita in the southern USA, all underscore the necessity for national and regional disaster early warning, planning and preparedness. Increased awareness of the need for dialog and coordinated disaster planning is also underway by the establishment of local and regional disaster management agencies such as the Indian National Disaster Management Authority, the Pacific Disaster Center (Laben, 2002) and by events such as the World Conference on Disaster Reduction (WCDR, 2005).

There is evidence that individuated interventions, without such coordination, can hamper effective response (Cyranoski, 2005). In recognition thereof, France, Europe and Canada signed a Charter on Cooperation for the purpose of coordinating their space-based technologies for aid 'in the event of natural or technological disasters' (Charter…, 2000). Similar cooperative spatial data resource deployment and associated research have been organized through NASA and the University of Maryland's Global Land Cover Facility (GLCF), the University of Alaska Fairbanks' Geophysical Institute, and the Multidisciplinary Center for Earthquake Engineering Research (MCEER, 2005). The data archive of Purdue University's Laboratory for Applications of Remote Sensing (LARS) and the near-real-time data from the Purdue Terrestrial Observatory (PTO) have been put on-line through the NSF TeraGrid and through the USGS-supported AmericaView program. Moreover, remote sensing applications have been developed for prediction of landslides (Mason and Rosenbaum, 2002), for assessing river basin sustainability (Jacobs, 1994), environmental management in watersheds (Gupta and Chen, 2001) and for addressing flood management (e.g. Flood Management…1998; Harris, 2006)

Ministries of Health, in collaboration with the United Nations' World Health Organization (WHO), and private sector organizations have coalesced to jointly address a wide array of insect-transmitted infectious diseases, mapping HIV/AIDS incidence and other sexually transmitted diseases (STD's) and their *sequellae*. Remote Sensing has been demonstrated to be a vital adjunct to conventional public health surveillance through monitoring infectious disease vector habitat, including the black fly *(Semulium damnosum)* vector for river blindness (onchocerciasis), the tse tse fly (*Glossina morsitans*) in the case of trypanosomiasis (Rogers and Randolph, 1993), the sand fly *(Phlebotomus paptasi & P. duboscl)* for leishmaniasis, *Anopheles gambiae* for malaria (Breman, et al., 2001) and the intermediate host snails, *Bulinus truncatus* and *Biomphalaria pheifferi & B. glabrata* for schistosomiasis. Similar monitoring utilizing remote sensing has been conducted for vectors of diseases which afflict feral and domesticated animal populations (Linthicum, 1999). Major emphasis has been devoted more recently to monitoring avian influenza, throughout Asia and Africa (Grose, 2004). Similar remote sensing – data fusion based applications can be found for plant pathogen transport, crop yields, and disease infestation. For example, Coops et al. (2006) and Wulder et al. (2006) show a data integration technique for remote sensing and *in situ* data to identify bettle infestation in forests.

Until fairly recently, remote sensing of urban areas was confined to 'urban heat island' effects. With the advent of higher spatial resolution data, the tools for analysis of urban landcover (Netzband & Stefanov, 2003) and landcover change at the urban fringe (Quarmby & Cushnie, 1989) have become more widely disseminated. The impact of urban sprawl on health and water availability as well as severe weather under changing climate is also now feasible with remote sensing data sets (Shepherd and Burian 2003; Pyle et al. 2008). Urban planners now make extensive use of GIS and in major cities and counties, establishing 'enterprise GIS' systems that are shared by tax assessors, fire departments, police departments, transportation authorities, utilities, voting districts, etc. The initial impetus for such systems was for cadastral mapping, which facilitated collection of *ad valorem* (i.e. property) taxes. The integration of remotely sensed data into urban enterprise GIS is occurring more slowly, as evidenced by the comparative paucity of papers on this issue at forums such as the Urban & Regional Information Systems Association (URISA) http://www.urisa.org.

A critical but underappreciated component of the disaster mitigation and recovery effort has its firm roots in environmental sustainability. This strategy of resilient development has been paved with successive movements initially focused on waste minimization, end-of-pipeline effluent reduction, pollution prevention, later, industrial ecology, product life cycle management and ultimately a more encompassing prudent intergenerational management of resources and long-term protection of the biosphere. The utility of remote sensing (Rochon et al., 2004) and Geographic Information Systems (GIS) (Blaschke, 1999) for attaining and maintaining environmental sustainability has also been postulated. The inherent interdisciplinary nature of remote sensing is evidenced by the wide variety of sectors to which it has been deployed.

For example, sustainable water resource strategies include watershed monitoring (Rochon et al., 2002), tropical hydrological change and riverine biodiversity in Asia (Dungeon, 2000), as well as developing the cyberinfrasturcture for interdisciplinary, multiscale studies for computation based discoveries (e.g. Govindaraju et al. 2008). The complex issues related to flood-recession agriculture during the dry season, particularly rice production, has been examined by Fox and Ledgerwood (1999). One of the early assessments of sustaining tropical forest resources in Thailand and Myanmar was conducted by the UN FAO (Lanly, 1982.) More recent discussions of deforestation and sustainability were offered by Vajpeyi (2001) and utilization of Landsat Thematic Mapper (TM) data for mangrove forest monitoring was presented by Kovacs, et al. (2001).

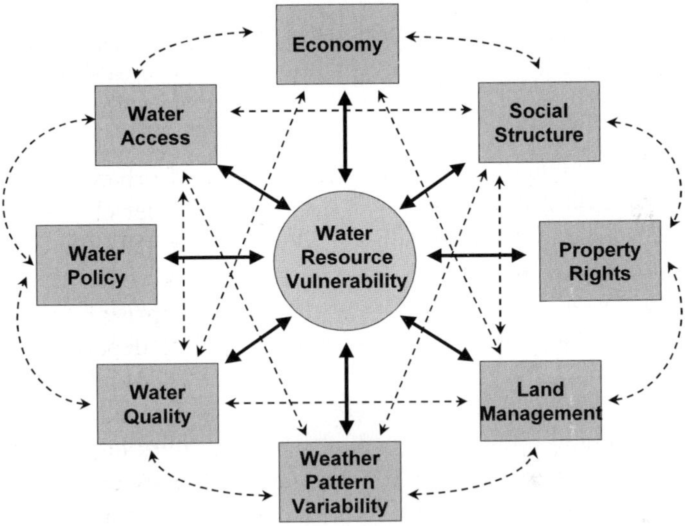

Fig. 1. Schematic linkages (solid arrows) and interactions (dashed arrows) related to changes in the vulnerability of water resources in India due changes in the Indian Monsoon (from Douglas et al. 2006)

Of paramount importance to the successful implementation of sustainable development best practices, is an intensification of emphasis on poverty alleviation- in which spatial technology is being integrated in various stages. Inter-generational management of natural resources can only be made palatable to impoverished communities, when sustainable development practices no longer impinge upon the survivability of the current generation and their immediate and extended families. The prototypes developed by the Thai Royal Family, recently acknowledged by an award from the UN, including those sustainable development initiatives associated with King Rama IX's Royal Projects and those represented by the Mae Fah Luang Foundation, wherein opium poppy growers were presented with economically beneficial crop substitution alternatives, could serve as a blueprint for replication elsewhere. 'Community involvement' in the planning and implementation process (Chenoweth et al. 2002) is one essential component of this approach to challenge the perceived conflict between biodiversity and poverty alleviation (Sanderson and Redford, 2003). There is even a role for Space Technology for Poverty Alleviation and Rural Development, as implied by the conference of that same name, organized by the United Nations Economic and Social Commission for Asia and the Pacific. However, before this becomes main stream, the Digital Divide, Bandwidth Divide and the Data Divide all need to be addressed in order to bring the promise of remotely sensed spatial data dissemination as a tool for sustainable development, poverty alleviation and disaster mitigation to full fruition (Data Sharing for Disasters, *Nature* Editorial, 2005).

Capturing the inter-relation of human actions, as well as natural processes, is a necessary but challenging task. An example of multiprocess/multisensor integration for environmental vulnerability mapping is shown in Douglas et al. (2006), who developed an irrigation and water use assessment for India, building on precipitation location and drought vulnerability. The Douglas et al. approach provides a conceptual synthesis to show that the mapping of agricultural landscapes and irrigation water use could potentially lead to changes in the projections of inter-annual variability in the monsoon rainfall, and perhaps more directly to changes in the intensity of inland rain events and water availability. A traditional approach adopted in identifying the vulnerability typically adopts a linear framework involving a singular forcing. A more interactive vulnerability assessment approach can be summarized for different resources as shown in Fig. 1. The interactions and the feedbacks are important pathways and they can lead to assigning different values to the landscape and their resilience needs (Pielke and Niyogi 2008).

5.3 Disaster Mitigation and Emergency Response

Disasters occur with regular frequency and while the community at large has adapted itself to these regular occurrences, the economic and social costs continue to mount year after year. Over one million houses are damaged annually in addition to other losses, including humans, livestock, and crops.

The Indian subcontinent, for instance, is vulnerable to natural disasters such as droughts, floods, cyclones, earthquakes, and landslides. Over half of the country is vulnerable to earthquakes, eight percent of the land is vulnerable to cyclones, and five percent of the land is vulnerable to flooding (UNDP India 2007). Also, there is a geographical demarcation for different hazard occurrences. Generally, the biggest earthquakes usually occur in north and northeastern states. Most cyclones occur along the east coast around the Bay of Bengal. Floods in the Indo-Gangetic-Brahmaputra plains are an annual feature. Most landslides occur in the mountain states along Himalayan foothills and the Western Ghats. Interestingly, almost the entire country can be vulnerable to drought because of monsoon variability or anthropogenic water usage beyond the regions carrying capacity. Also, because of the summer monsoon, the majority of natural disasters occur between the months of May through November. There are two cyclone seasons, May – June and October – November. Additionally, cyclones occur along the coastal states neighboring Bay of Bengal in September and in southern states typically in December. Floods usually occur during the months of June through October when the monsoon trough is at its maximum extent over the Indian subcontinent.

Meteorological departments in different regions broadcast early warning to the Government at the central and state levels. Additionally, early warning messages are broadcast through print and electronic media. Emergency response is mainly a state function and a Special Relief Commissioner is solely in charge of the entire operation. Emergency response methods include early warning dissemination, evacuation, search and rescue, relief, and rehabilitation assistance. Depending on the nature and degree of the disaster, armed and para-military forces are also requested by the state government.

Many disaster management policies and programs have been developed and implemented in the last several years in India under the National Committee on Disaster Management (NCDM), aimed at reducing adverse impacts and fatalities from disasters through consistent monitoring, quick and efficient response (Sinha, 2003)

While all efforts and components of such disaster management programs are very important, especially in the areas of spatial analysis involving modeling, visualization and fusion of archival, real-time multi-sensor satellite data, socio-economic data and in-situ data, the setup of early warning systems (EWS) to predict potential degradation trends and disasters early in time before their actual occurrence would allow for better planning and emergency response action to help save more lives and properties.

An EWS is an integrated system for monitoring, data collecting, analyzing, interpreting, and communicating monitoring data, which can then be used to make decisions early enough to protect public health and minimize unnecessary concern and inconvenience to the public. Depending on the type of EWS, all relevant institutions and resources at government, state, county and local levels should collaborate in the initial setup of any EWS and should equally be furnished with important advance warning information for efficient implementation of emergency response plans. A well integrated EWS is expected to have four major tech based components in the areas of:

- Information and data collection on the phenomenon
- Evaluation, processing and analysis of collected data
- Dissemination of warning information to policy makers and end-users
- Implementation of an effective and timely response to the early warnings issued

A case study was performed on the flood that occurred in September, 2000 in Gangetic West Bengal to identify the rural settlements that are vulnerable to floods of a given magnitude. In locating the vulnerable areas, two factors are observed: the presence of deep flood water in and around the settlement and its proximity to an elevated area for temporary shelter during an extreme hydrological event. Landsat ETM images from September 30, 2000, ASTER digital elevation data, and ERS-1 synthetic aperture radar data were used to identify non-flooded areas within the flood zone, to assess and rectify the classified image, and to extract the settlement areas surrounded by trees, respectively. Once the satellite imageries were extracted, they were imported into ArcGIS and then spatial analysis, using socio-economic data, was carried out to identify the settlements vulnerable to river inundation.

The study provided a reasonably accurate spatial information database that is appropriate for generating 1:250,000 hazard maps. However, there are inherent errors in classifying flooded areas from non-flooded areas as a result of cloud cover over the area and the predominance of tree canopy in the rural area. Also, the lack of high resolution digital terrain data led to poor accuracy of the classification results. Additionally, the lack of an updated large-scale map restricted the ability to carry out a detailed mapping of the study area.

The spatial and temporal characteristics of most of the natural disasters in India require sensor techniques for monitoring and data collection that allow for all trends at specific times to be monitored. Most current technologies include earth observation (EO) technologies such as multi–sensor remotely sensed data as well as from photogrammetry with varied resolutions. The application synergy between low and high resolution sensors and real-time and archival data allow for the combination of different spatial, temporal, spectral and radiometric resolutions to i) identify trends and locate vulnerable areas (small scale monitoring) and ii) assess vulnerability and predict possible scenarios (large scale mapping) (Holecz et al., 2003).

Most commonly used coarse resolution satellite systems for regional land degradation related environmental and climatic monitoring include the NOAA AVHRR, MODIS, ENVISAT, METEOSAT, and GOES, most with high temporal resolutions.

During periods of land degradation and associated drought conditions, physiognomic changes and differences within vegetation can be measured drought indices (e.g. Sims et al. 2001) and by the manipulation of satellite measured spectral radiance into useful products and indexes (Tucker and Choudhury, 1997) which are then used as proxies for the estimation of vegetation health, soil water content and stress (Kogan 1995).

Commonly derived indexes and products include Normalized Differential Vegetation Index (NDVI), Green Vegetation Fraction (GVF), Enhanced Vegetation Index (EVI), Soil Adjusted Vegetation Index (SAVI), Moisture Stress Index (MSI), Leaf Area index (LAI), Leaf Water Content (LWC), Net Primary Production (NPP), Rain Use Efficiency (RUE), Local NPP Scaling (LNS), Photosynthetically Active Radiation (PAR), Gross Primary Production (GPP) and Fraction of PAR (fPAR) (Prince and Goward 1995). The recent availability of AMSR and GRACE datasets with capabilities for soil wetness estimation and large regional scale soil moisture balance also provide useful indicators for drought, flooding, and landslide potentials.

For the purposes of monitoring short-term events such as environmental pollution and geologic activities (e.g. landslides and volcanic activities), multi purpose field sensors, weather stations and global positioning systems (GPS) are utilized for *in-situ* data collection. The integration of all such collected data, products and indexes, together with socio-economic data such as population growth, market trends, food availability and prices, health and malnutrition, industries and agriculture serves as important input component for EWSs (Holecz et al., 2003).

After initial data evaluation and integration, environmental modeling, GIS, statistical and probabilistic analysis are done to establish or predict trends and intensity of a degradation phenomenon or threatening climatic conditions, using established baseline for degradation threat and the level of certainty of the EWS. Remote sensing products now have a history of over two decades in terms of landcover datasets and the more recent products have significant high resolutions, both temporally and spatially, to develop robust statistical relations. Using GIS and visualization analysis, risk assessment maps are produced detailing areas for people who would be most vulnerable to a potential upcoming disaster. This serves as the basis for issuing necessary warning information.

Deviations from established baseline standards for a particular degradation or disaster triggers a chain of communications (via internet, satellite, wire, wireless, etc.) which informs all relevant authorities and decision makers (i.e. politicians, law enforcement agencies, medical expert and engineers) at the national, state and local levels responsible for various aspects of emergency response. Based on their immediate decisions, already developed emergency response plans such as evacuations, medical treatment, etc. are theoretically implemented to protect the public and the environment from negative impacts.

Warning information, such as that for drought and famine EWSs, identify drought affected areas and help provide good estimates on crop yield for the region. This is important for planning, seeking funds and budgeting to provide food supplements and other essential needs for such affected regions in India to help avert hunger and suffering. An example of the need for EWSs, according to the WMO was the reduced death toll of 200 compared to 130,000 from similar cyclones in 1991 and 1994, respectively, in Bangladesh and eastern India, as a result of the advanced warning for the more recent cyclones (e.g. De et al. 2005).

While critical for disaster management and human survival, establishing well integrated and reliable Early Warning System (EWSs) requires overcoming numerous challenges. Issues related to funding, research, standardization, expertise, sound technologies (e.g. sensor designs), infrastructure, institutional capacity, collaboration and integration, efficiency in response, improved risk management, as well as communication, need be looked at holistically in the process of setting up EWSs. Since emergency response is the single factor determining the final impact of most disasters, it is very crucial that emergency response plans such as evacuations, food and medical aid be made in advance, rehearsed and ready to be implemented for a wide array of potential disasters. In addition, the public in general needs to be aware of what is expected of them and have trust in the process. The best plans will not be effective if those being affected are not willing to follow and participate/help in the emergency response procedures.

5.3.1 Archival and Real-Time Remote Sensing

Sources for global and regional-scale remote sensing data and derived products are continually expanding. For example, archival MODIS data are available from the National Aeronautical and Space Administration (NASA); although one needs to register to obtain some of the products (http://www.modis.gsfc.nasa.gov). Also, some Landsat MSS and TM images are available for almost any land area of the globe at the Global Land Cover Facility (GLCF) at the University of Maryland. Other remote sensing image data available at the GLCF are ASTER and Shuttle Radar Topography Mission (SRTM) data. Additionally, several products derived from MODIS and AVHRR data are also available including MODIS 32-day Global Composites, MODIS 16-day NDVI composites, MODIS Vegetation Continuous Fields, AVHRR Global Land Product and AVHRR Continuous Fields Tree Cover. For the tropical region, the Tropical Rainfall Measurement Mission (TRMM) based microwave imager (TMI) data for rainfall amounts, soil moisture, and convection are also important from improving the disaster management potential. Recent studies integrating satellite products into high resolution regional weather models continue showing high degree of applicability for the development accurate forecasts for tropical storm situations (Xavier et al. 2008, Vinodkumar et al. 2008).

Near real-time MODIS data are available from the University of Wisconsin (http://eosdb.ssec.wisc.edu/modisdirect/) for the coverage area of their tracking antenna, and from Oregon State University (http://picasso.coas.oregonstate.edu/ORSOO/MODIS/DB/). Data is available from the Purdue Terrestrial Observatory (http://www.itap.purdue.edu/pto/) after mid 2006 from the NOAA AVHRR, NOAA GOES, Terra MODIS, Aqua MODIS and Feng Yun 1D MVISR sensors.

Another model of promoting relatively easy access to remote sensing data is being provided by AmericaView (http://www.americaview.org/), a USGS sponsored activity. There are currently 33 state-views which individually provide public domain image data for their state areas. The purpose of AmericaView is to promote the use of remote sensing in the university, governmental and private sectors via workshops and portals to the remote sensing data. To date, most of the data online includes freely-accessible Landsat Thematic Mapper (TM) data for the individual states. In the United States, an important resource for data dissemination is based on a model involving national dissemination by the National Climatic Data Center, at regional scale by the regional climate centers, and at the state level by the state climate offices. For the state of Indiana, for instance, a statewide data access and archival center – Indiana State Climate Office (Iclimate.org) is located at Purdue University. These resources, in concert with various other state, federal and private partnerships provide redundant and, therefore, highly accessible information under pre and post disaster conditions.

From the Indian disaster mitigation perspective, development of a state climate office – possibly in conjunction with the India Meteorological Department and academic or national lab partnerships, as appropriate, can provide an alternate avenue for disaster planning mitigation and response. A common cyber gateway that can provide access to the different products and models to integrate these datasets into a format that can be easily linked within models for research, education, and secure operations is still lacking. Such an interface would likely be developed targeting specific disaster in mind along with flexibility for adopting future technological developments.

As an example, the heavy rain events are expected to increase across the world in response to the changing climate (IPCC, 2007). The response for such events would require high resolution rainfall monitoring *(in situ* and remotely sensed) such as is expected from the Global Precipitation Mission and the current Tropical Rainfall Measurement Mission (TRMM) sources. The US NOAA Climate Prediction Center (CPC) Merged Online Precipitation Data (CMORPH) is one such interface that is currently available. However, for accurate prediction, a range of model inputs are needed – from real time vegetation state to accurately represent the transpiration moisture feedback from the land surface (e.g. from MODIS vegetation greenness fraction or the AVHRR land use categorization); the sea surface temperature for resolving any energetic eddy that has not been resolved in the model ocean fields (such as from TRMM, AMSR, or SSM/I fields), the surface soil moisture at fine and regional scales to ascertain the evaporation and runoff feedback from the bare soil surface (such as from AMSR E soil moisture fields and the GRACE water balance estimates, and the upcoming SMOS- Soil Moisture and Ocean Salinity satellite sensor), the cloud motion and thickness to resolve the cloud structure and the rainfall potential (such as from Cloudsat and the CALIPSO as well as the Quickscat cloud motion vectors), the vertical sounding of the atmosphere such as from NOAA – TOVS or the COSMIC datasets- for developing better vertical structure and initial conditions/ assimilation within the model; the atmospheric radiative balance such as from ERBE fields, etc. Clearly, the challenge would be in identifying the minimum requirements first and then developing the scalability for a range of possible scenarios, such a multisensory assimilation facility, would need to cater to. As outlined in Pielke et al. (2007), the potential currently exists that one could develop a series of the model parameterizations that can provide simplified cloud models that only rely on remote sensing sources. These remote sensing based model estimates can be integrated within the current and near future remote sensing data, as well as more complex models for quicker analysis. Such a series of multiple simpler models could potentially provide an interface for developing scalable applications from the multisensor datasets for disaster management.

5.4 Dynamic modeling integration and data fusion

An important pathway for an integrated framework for the geospatial data into the disaster mitigation and planning tool is to develop dynamic modeling interfaces between multisensor/ multiscale datasets.

As an example, Kumar et al. (2008) combined *in situ* rain gauge and the Tropical Rainfall Measurement Mission (TRMM) PR (precipitation radar) data with multiple models and dynamical parameterizations to diagnose the development and the uncertainty in the simulation of the record heavy rain event that occurred over Mumbai on July 26, 2005. The heavy rain caused unprecedented damage over Mumbai for a single day rain event. In less than 24-hours, nearly 750 mm of rain fell over northern Mumbai and a location about 27 km away in the southern part of Mumbai, received about 30 mm. Most of the operational centers were unable to predict this record-breaking rain event, which eventually was responsible for over 400 deaths and nearly a billion USD loss.

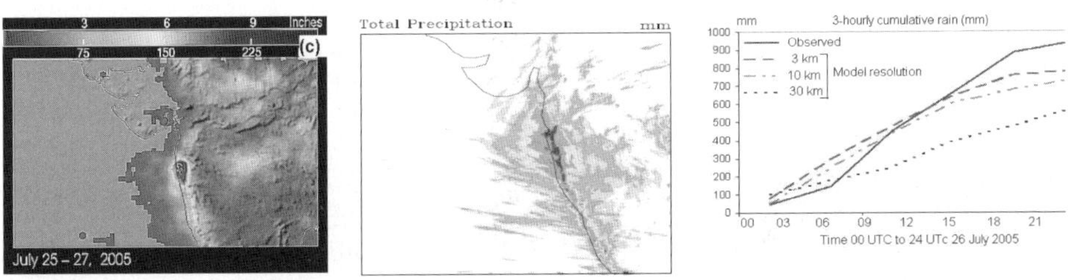

Fig. 2. TRMM satellite derived (left panel) and WRF Model simulated 24-h spatial and temporal rainfall for July 26, 2005 (central and right panel) for different model grid spacing. Inclusion of high resolution features and calibration with remote sensing products helped improve the performance (from Kumar et al. 2008; and Chang et al. 2008).

Using high resolution model grid spacing (~3km versus 50km operationally), and integration of a number of *in situ* and remotely sensed products, the weather research and forecasting (WRF) model was able to simulate a very localized heavy precipitation event (as shown in the Fig. 2 Chang et al. 2008). Kumar et al. (2008) conclude that for simulating such events over a megacity in the future there is a need for developing a localized Doppler radar monitoring capability, and ingestion of the radar and satellite products such as from the TRMM and MODIS (e.g. sea surface temperature, and rain rates) into the dynamical models.

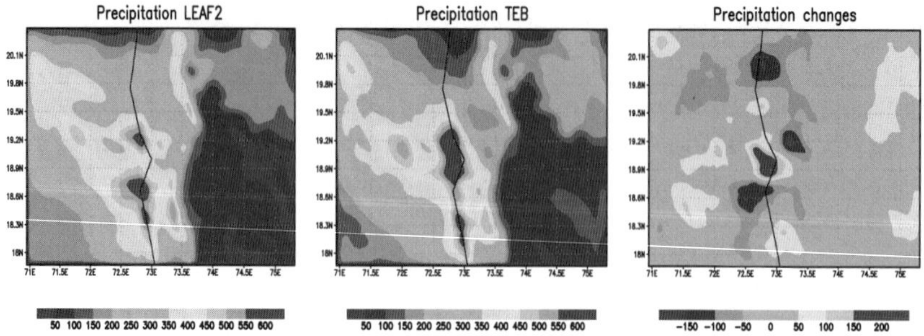

Fig. 3. Default model rainfall using a 1 km model grid (left panel) and the satellite data assimilated rainfall (central panel). The different is shown in the right panel (Lei et al. 2008)

The impact of introducing the Mumbai urban area surface characteristics explicitly using satellite data sets such as MODIS and AVHRR, and prescribing near real-time sea surface temperature gradients using remote sensing data sources such as from the SSM/I dataset (instead of using the default climatology) are shown in Fig. 3.

5.5 Visualization

With the access to data from multiple sources, it is not uncommon to have spatial data sets with terabytes of data. However, the technology to process and display these kinds of data in real-time is still lagging. Traditional high-end systems to handle such large and complex data sets are very expensive.

A GIS system allows one to manage, analyze, and display geographic information which is represented as layers of information (ESRI-1, 2004). This information set includes – maps, globes, geographic data sets, processing and work flow models, data models, and metadata. The broad area of GIS can be viewed in different ways. For example (ESRI-2):(a)as a spatial database of features, rasters, topologies, networks, etc., (b) as a set of intelligent maps with several views that shows features and feature relationships on the earth's surface to support queries, analysis, and editing of the information, or (c) as a set of tools and functions to transform information that derives new geographic data sets from existing datasets. It is obvious that display of information is an important part of any GIS system. Large maps that are opened and spread flat give much more information than when they are seen folded. The more information one can see without loosing finer details, the better we can detect and identify objects and the better can one make associations and connections between various features on a map. The human eye has one of the best image processor. One can make faster and better decisions than computers in many situations. This simple philosophy drives one to find better ways to visualize geographic data.

To process and visualize large data sets effectively requires combination of high-performance computing, smarter software algorithms, and large display systems. Traditionally, desktop based graphics rendering is performed using a single personal computer (PC) on a single monitor. With the latest developments in graphics hardware and computing power, a single PC is used to drive multiple displays at the same time. But the complexity of scientific visualization applications increased tremendously, so that a single PC could not render complex scenes in real-time even to a single display, not to mention multiple displays. This single PC, in single and multi display configurations, has several limitations, such as insufficient computing power, trade-off between rendering quality vs. display area, especially to render complex scenes in real-time. To overcome these limitations, a commodity cluster was deployed to render complex scenes to a single monitor, and/or to different tiled display configurations in real-time. This approach provides more flexibility and ability to customize the computing power, rendering quality and display area as required.

Fig. 4. Three dimensional map rendition of flooding event over Lafayatte, IN at Purdue University.

While the above systems provide high-performance computing, rendering and display area, most of the commercially available software cannot be used 'as is' on these specialized hardware configurations to its maximum potential. Most of the current software applications are limited and focused on solving specific problems. The objective was to be able to run visualization software on these high-performance systems, seamlessly, so that the user is able to visualize with the same ease as if it were running on a desktop workstation. The applications are in 3D and stereoscopic, and because they are rendered by a very powerful graphics cluster these displays are navigable in real-time. This software application not only takes advantage of distributed computing but also the distributed rendering for large, very high resolution display systems.

The image rendered in Fig. 4 is a visual result of a flood simulation. This flood simulation was developed using 1 foot color ortho images draped over a digital elevation model provided by 5-foot resolution LIDAR data. The data were provided by the Texas Advanced Computing Center (TACC) as part of a Flood Modeling demonstration to illustrate the potential of TerraGrid type activities for facilitating real-time simulation of emergency response personnel and evacuation planning for flood events over large regions. This was a joint effort involving TACC, DOE Oakridge National Research Labs (ORNL) and Purdue University.

Fig. 5. Aerial visualization of Greater Lafayette, Indiana

Figure 5 is visualization of the greater Lafayette region in Indiana, USA. The model shown consists of two sources of data: a 1m resolution color digital aerial photograph and a 30m resolution digital elevation model (DEM) of the area. ArcGIS software was used to drape the color image onto the DEM. Then the 3D draped model was then converted to VRML format and then to 3DS Max™ file format. This 3DS file was eventually used to display the model using inbuilt cluster aware GIS rendering application on the tiled wall display.

The above figure also shows the display of multiple layers of information of the Purdue University Campus. They include – a one foot resolution aerial imagery of the Purdue University Campus, a layer of information with buildings, roads, sidewalks, and different species of trees, and a 1m resolution LIDAR data. The technique used to visualize this project was the same as discussed in previous application. Each layer was visualized in the

ArcGIS 3D environment then passed to the cluster-aware GIS rendering application on the tiled wall display.

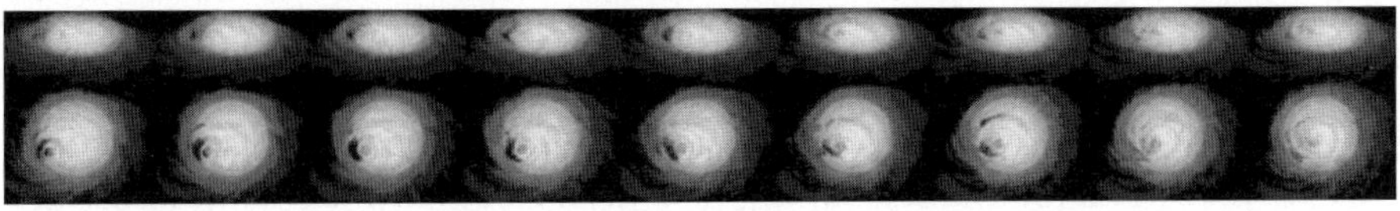

Fig. 6. Time varying Doppler Radar data of Hurricane Katrina (2005)

Figure 6 shows time sequential Doppler radar data of Hurricane Katrina that made landfall and caused unprecedented damage to New Orleans, Louisiana in 2005. The images are rendered in both top and front view.

The above described applications demonstrate the usefulness of visualizing complex data sets on large display systems. This gives us a very high resolution display that shows much more detailed information. While the software applications for visualizing individual aspects of these data sets are being developed, integrating all these data sets into a single system that coherently interrelates and provides an effective visualization interface for the users is still one of the biggest challenges.

5.6 Wireless Technologies / Handheld Computers

The use of advanced information technologies that can deliver critical information anytime, anywhere will play a critical role in disaster management and mitigation. Although this area of work is still in its infancy, cybertools that are developed under this umbrella of activity will have the ability to deliver just-in-time intelligence for appropriate action by decision makers. This area of work is so critical and cutting-edge that the U.S. National Oceanic and Atmospheric Administration (NOAA) has included this as a key part of its strategy (http://www.cio.noaa.gov/hpcc/crisis.html). This effort in the United States has received support from multiple sources including the Federal Emergency Management Agency (FEMA).

The effort to provide mobile tools includes and is innately informed by another area of study – namely location and context awareness of analytic tools. Context-aware computing is a large area of research that attempts to provide users with performance, tools, and interfaces optimized for specific scenarios of use. Modern handheld devices not only carry significant computation capabilities that allow researchers to build, deploy, and prototype such next-generational tools, they also carry on-board sensors that can help detect various environmental parameters which, in turn, can be relayed back to base stations that are away from the disaster zones. A critical number of first responders carrying such devices could easily form ad-hoc sensor networks that can capture significant amount of data, which as part of an integrated workflow, can be fed to computational models to forecast potential hazards or provide decision support. For example, the use of RFID technologies, in combination with sensor chips, are being used extremely successfully by large industrial units that provide logistical and supply chain support. This in combination with commodity services like Google Earth (http://earth.google.com), that utilize a service-oriented paradigm can easily be delivered to cell phones or other mobile computational units.

Another area of work that has tremendous potential for the customization of services to mobile computing environments is the use of virtual machine technologies. Virtual machines can be intrinsically designed to deliver full-fledged computational services that resemble full workstations. The added advantage of virtual machines is that they are designed to migrate and pull data across network domains. Significant progress is made on this front with support from the National Science Foundation's National Middleware Initiative.

To summarize, the use of mobile technologies is still fairly new. However, given the ability of mobile computing platforms to engage significant computational power, coupled with advances in RFID, Bluetooth (extremely short-range communication), and cellular network bandwidth, it is only a matter of time before serious research and production grade software is released for disaster management, mitigation, and response-use case scenarios.

5.7 Education and Training

Educating the next generation of scholars, leaders and policy makers in remote sensing – model data fusion using remote sensing techniques will require a curriculum-wide integration, analogous to the migration of 'ethics,' previously the exclusive province of Philosophy departments, within a wide array of academic departments, including medicine, business, journalism, pharmacy, political science, engineering, etc. (Rochon, et al., 2006).

The emergence of integrated approaches to disaster management and mitigation has exposed the critical need for new and innovative strategies for training not only the next generation of first responders, but also for disaster management teams, scientists who can predict such catastrophic events, and decision makers. One of the most critical characteristics that need to be accounted for is that any education and training in this area needs to resemble and mimic closely the real world. Authenticity of the learning experience that allows learners to solve complex problems and use hands-on techniques will be valuable in ensuring transfer of skills from the educational setting to real world contexts. There is a need for utilizing 3D simulations and visualizations as an intrinsic part of the training process for handling disaster management scenarios. The field of disaster management and mitigation can be considered the larger problem space to which this approach of using 3D simulations and immersive environments can be used. However, the availability of such advanced and expensive environments in many parts of the world is limited and we have to look at lower cost ways for delivering similar training and educational opportunities.

The use of simulations has been long considered a successful strategy for incorporating near real-world scenarios into the educational curricula (Finkelstein 2005, Norman and Spohrer 1996). While the high-end of simulation use for training requires the availability of complex hardware tools, there are efforts such as the U.S. National Science Foundation funded Network for Computational Nanotechnology (NCN). The front-facing side of the NCN is a science gateway called the nanoHUB (http://www.nanohub.org), which provides users with easy access to advanced, industrial strength simulation tools that also tie into the national cyberinfrastructure fabric. The use of such science gateways for disaster management and mitigation could be invaluable to not only learners, but also to scientists. The emergence of science gateways to deliver advanced simulation for research and education has also highlighted the need for communities to have a single point of computation, data, and tool access. The ability to collaborate and contribute community approved and reviewed code to bridge education and research will be a key ingredient that will determine the success of disaster management and mitigation efforts.

In the context of developing countries such as India, the technology infrastructure to deliver training that utilizes tremendous amounts of data and advanced visualization tools that rely on heavy computational environments is still evolving. The use of science gateways as an aggregation point for the various data and analytic tools is not only feasible, but is also a practical, viable decision at this time point. Such efforts not only lower entry barriers to the field, but also provide educators with continuous engagement to the science and the scientific community. Similar efforts to train educators in the computational sciences in the United States deployed as part of the IEEE Supercomputing conference's education program – require educators to engage with materials related to the application of as part of a year-long program. The program structure allows participants to use web portals that utilize the open source course management ***Sakai*** to learn, practice various cutting-edge techniques, and for assessment. Participants also receive significant face-to-face interaction with expert instructors. The core of all such efforts is the use of science gateways that allow research materials to be disseminated directly and efficiently to the educators.

The area of disaster management and mitigation requires education and training efforts to 'think outside the box.' Efforts to reach beyond traditional classroom paradigms to tap rich data and computational tools are key to the future strength of such initiatives. Additionally, education and training programs have to reach for resources well outside the physical confines of a single environment. Aggregation and prioritization education and training around science gateways that provide access to simulation tools and lower entry barriers are strategically essential.

5.8 Conclusion

Mitigating disasters is a complex undertaking that takes the expertise of many different disciplines - geologists, hydrologists, meteorologists, engineers, the medical community, policy makers, and communication specialists. Moreover disaster mitigation requires the use of many different types of data and tools - including remotely sensed images and point sensor measurements, GIS, wireless technologies, visualization applications, physical models of areas, mathematical models. Disaster mitigation involves different time scales for planning: longer time scales when planning for climate changes, earthquakes; shorter time scales for seasonal events such as fires, flooding, hurricanes. Implementation of disaster mitigation, though, after an event has a very short time scale. In many cases, lost hours and minutes could mean lost lives. In these cases, one cannot make the plan up as one goes. A plan with evacuation or response procedures must be in place and understood by all in the affected area, trained personnel must be available, the information required for making decisions by those in charge of mitigation must be readily available, such as up-to-date images of the affected area, utility availability, rescue equipment availability, awareness of potential toxic sources in the area, to name just a few.

Dynamical models can be important tools for integrating the multiscale / multisensor data and utilize the heterogeneous information for disaster mitigation and emergency response. The increasing access to satellite datasets for multiple variables and scales makes adaptation of remote sensing technology an exciting tool. The challenges lie in developing a common, readily available interface for querying the different data and running the models in a mutliscale framework. The inherent limitations in the different environmental models for resolving multiple spatial scales will likely be a significant problem in the near future. The ability to assimilateremote sensing data to help represent many of the processes that are poorly simulated in the models continues to show increasing benefits to the disaster mitigation and management efforts. The subsurface phenomenon, such as that due to ground water seepage, or the processes involving phase change (such as runoff due to snow melt and subsequent evaporation leading to increased convection); or the processes within an urban area or under dense forest canopy for understory processes would likely continue being important challenges in this endeavor (for adopting remotely sensed data into disaster management tools).

There is also a paucity of simpler models that can provide first order approximations with rapid response, in place of slower but highly detailed process-scale models. Approaches involving look-up tables or diagnostic techniques, primarily based on remotely sensed parameters and approximations, are possible, but as yet under utilized in satellite data assimilation within disaster management technology.

The promise of wireless devices becoming exponentially more computationally powerful makes them an attractive option. The accessibility to this technology for mass implementation and the robustness for ready availability under disaster response mode would be a challenge the scientific community will work with in the coming years.

The future efforts for developing useful applications involving multiple datasets and data fusion techniques in decision making algorithms would likely not be limited by the availability of high resolution, multiscale satellite datasets, but by the limitations to the cyberinfrastructure, with respect to bandwidth, storage, and processing speeds available for developing rapid response applications capable of reviewing and managing different scenarios. The latter requires a community response that will require the cooperation and collaboration of governments, industry, academia and international organizations, that can provide the different components of such a cyber-enabled collaboratory. While such efforts are underway in a research scale through target programs from agencies such as the United States National Science Foundation, the efforts are largely targeted for environmental observatories (e.g. Govindaraju et al. 2008). Therefore, a broader response from the international community is necessary for developing global applications and disaster management collaboratories. Remote sensing, multiscale data fusion, satellite data assimilation, visualization and hierarchy of simple to more complex modeling approaches will likely play significant roles in defining the scope and scale.

References

Blaschke T, 1999, Sustainability with GIS: http://www.sbg.ac.at/ geo/people/ tblaschke/ publications/ sustain_abs.html.

Breman JG, Alilio MS, and Mills A, 2001, The Intolerable Burden of Malaria: A New Look at the Numbers, American Journal of Tropical Medicine and Hygiene 64 (1, 2 Suppl): 1 106.

Chang Hsin-I, A. Kumar, D. Niyogi, U. C. Mohanty, F. Chen, and J. Dudhia, 2008, Impact of convection and land surface parameterizations on the simulation of the July 26, 2005 heavy rain event over Mumbai, India. Global Planetary Change, accepted.

Charter on Cooperation to Achieve the Coordinated Use of Space Facilities in the Event of Natural or Technological Disasters, 2000, European Space Agency (ESA), French Space Agency (CNES), Canadian Space Agency (CSA) Rev. 3.2 (May 25, 2000).

Chenoweth JL, Sarah A, and Bird E and JF, 2002, Procedures for Ensuring Community Involvement in Multi-jurisdictional River Basins: A Comparison of the Murray-Darling and Mekong River Basins, Environmental Management, 29: 497-509.

Coops NC., Wulder MA, and JC White, 2006, Integrating remotely sensed and ancillary data sources to characterize a mountain pine beetle infestation, Remote Sensing of Environment, 105, 83-97.

Cyranoski D, 2005, Solo Efforts Hamper Tsunami Warning, Nature, 433, Jan. 27, 2005, p. 343.

Data Sharing for Disasters, 2005, Editorial, Nature, 433, 271, January, 2005, p. 339.

De US, RK Dube, GS Prakasa Rao, 2005, Extreme weather events over India in the last 100 years, J. Ind. Geophys. Union, 9, 173 – 187.

Douglas E, Niyogi D, Frolking S, Yeluripati JB, Pielke, RA Sr., Niyogi N, Vörösmarty CJ, Mohanty UC, 2006, Changes in moisture and energy fluxes due to agricultural land use and irrigation in the Indian Monsoon belt, Geophysical Research Letters, 33, L14403, doi:10.1029/2006GL026550.

Dungeon D, 2000, Large–Scale Hydrological Changes in Tropical Asia: Prospects for Riverine Biodiversity. Cambridge University Press, Cambridge, UK.

Flood Management and Mitigation in the Mekong River Basin, 1998, FAO of the UN. Proceedings of the Regional Workshop. Vientiane, Lao PDR, 19-21 March. RAP Publication # 1999/14. http://www.fao.org/docrep/004/ac146e/ AC146E00.htm#TOC.

Filippidis A, Jain LC, and Martin NM, 1999, Using Genetic Algorithms and Neural Networks for Surface Land Mine Detection. IEEE Transactions on Signal Processing, 47 (1): 176-186.

Finkelstein ND, Adams WK, Kller CJ, Kohl PB, Perkins KK,, Podolefsky S, Reid S., Lemaster R., 2005, When learning about the real world is better done virtually: A study of substituting computer simulations for laboratory equipment, Phys. Rev. ST Phys. Educ. Res., 1, 010103, 8 pages.

Fox J, Ledgerwood J, 1999, Dry-Season Flood-Recession Rice in the Mekong Delta: Two Thousand Years of Sustainable Agriculture? Asian Perspectives: the Journal of Archaeology for Asia and the Pacific, 38.

Global Land Cover Facility (GLCF). Tsunami + Remote Sensing + Protected Areas. NASA and University of Maryland, Dept. of Geography & Institute for Advanced Computer Studies. http://glcf.umiacs.umd.edu/portal/tsunami/.

Govindaraju , S., et al. 2008, A cyberinfrastructure for environmental exploration, engagement, and education (C4E4), ASCE J. Hydrol. Engg (Cyberinfrastructure for Environmental Observations- Special issue), in press.

Grose C MD, 2004, Avian Influenza Virus Infection of Children in Vietnam and Thailand. Pediatric Infectious Disease Journal. 23: 793-794.

Gupta A, Chen P, 2001, Remote Sensing and Environmental Management in the Mekong Basin, 22nd Asian Conference on Remote Sensing, Singapore, 5-9 November.

Habib MK, 2001, Mine Detection and Sensing Technologies-New Development Potentials in the Context of Humanitarian Demining. IECON'01 27th Annual Conference of the IEEE Industrial Electronics Society, 1612-1621.

Harris P, 2006, NGIS-Australia: Mekong River Commission: Remote Sensing & Flood Management. http://www.ngis.com.au/Success+Stories/International+Projects/ 59.aspx.

Holecz F, Heimo C, Moreno J, Goussard J, Fernandez D, Luis Rubio J, Erxue C, Magsar E, Lo M, Chemini A, Stoessel F, and Rosenqvist A, 2003, Desertification – A land degradation support service. IEEE 3:1490 – 1492.

Italian Cooperation & UNCCD, 1999. Early warning systems and desertification, pp. 30.

IPCC, 2007: The intergovernmental panel on climate change – www.ipcc.ch accessed November 2007.

Jacobs JW 1994, Toward Sustainability in Lower Mekong River Basin Development. *Water International,* 19, 43-51.

Kogan FN, 1995, Droughts of the Late 1980s in the United States as Derived from NOAA Polar-Orbiting Satellite Data. Bull. Amer. Meteor. Soc., 76, 655–668.

Kovacs JM, Wang J, and Blanco-Correa M, 2001, Mapping disturbances in a mangrove forest using multi-date Landsat TM imagery. Environmental Management, 27: 763-776.

Kumar A, Dudhia J, Rotunno R, Weismann M, Niyogi D, and Mohanty UC, 2008, Analysis of the 26 July 2005 Heavy Rain event over Mumbai, Quart. J. Roy. Meteorol. Soc., in review.

Laben C, 2002, Integration of Remote Sensing Data and Geographic Information System Technology for Emergency Managers and their Applications at the Pacific Disaster Center, Optical Engineering, 41: 2129-2136.

Lanly J, 1982, Tropical forest resources. Food and Agriculture Organization of the United Nations (FAO) paper no. 30. Rome: FAO, UN.

Lantieri, D, 2003, Potential use of satellite remote sensing for land degradation assessment in drylands. Published by LADA, pp. 77.

Lei M., D. Niyogi, C. Kishtawal, R. Pielke Sr., A. Beltrán-Przekurat, T. Nobis, and S. Vaidya, 2008, Effect of explicit urban land surface representation on

the simulation of the 26 July 2005 heavy rain event over Mumbai, India, Atmos. Chem. Phys. Discussions, accepted.
Linthicum, KJ, 1999, Climate & Satellite indicators to forecast Rift Valley Fever epidemics in Kenya. Science 285, 397-4000.
Mason PJ, Rosenbaum MS, 2002, Geohazard mapping for predicting landslides: an example from the Langhe Hills in Piemonte, NW Italy. Quarterly Journal of Engineering Geology & Hydrogeology 35: n317-326; DOI: 10.1144/1470-9236/00047.
MCEER, 2005, Multidisciplinary Center for Earthquake Engineering Research (MCEER), Remote Sensing Technologies Applied Following South Asian Tsunami, mceer.buffalo.edu/research/ tsunami/ default.asp.
Netzband M, Stefanov WL, 2004, Urban Land Cover and Spatial Variation Using ASTER and MODIS Satellite Image Data. ISPRS Commission VII, Istanbul, Turkey.
Norman DA, Spohrer JC, 1996, Learner centered education, Communicaitons of the ACM, 39, 24 – 27.
Pielke Sr R.A., Stokowski D, Wang J.-W., Vukicevic T., Leoncini G., Matsui T, Castro C., Niyogi D., Kishtawal CM, Biazar A., Doty K., McNider RT, Nair US, Tao EK, 2007: Satellite-based model parameterization of diabatic heating. EOS, 88, 96 - 97
Pielke Sr. R.A., and D. Niyogi, 2008, The role of landscape processes within the climate system. In: Otto, J.C. and R. Dikaum, Eds., Landform - Structure, Evolution, Process Control: Proceedings of the International Symposium on Landforms organised by the Research Training Group 437. Lecture Notes in Earth Sciences, Springer, Vol. 115, in press.
Prince SD, SN Goward, 1995, Global primary production: A remote sensing approach, J. Biogeography, 22, 815 – 835, doi:10.2307/2845983
Pyle P., D. Niyogi, S. P. Arya, M. Shepherd, F. Chen, B. Wolfe, 2008, An Observational and Modeling-based Storm Climatology Assessment for the Indianapolis, urban region, J. Appl. Meteorol. and Clim., in revision.
Quarmby N, Cushnie J, 1989, Monitoring urban land cover changes at the urban fringe from SPOT HRV imagery in south-east England, International Journal of Remote Sensing 10: 593-963.
Reichle RH, Koster RD, Dong J, Berg A.A., 2004, Global Soil Moisture from Satellite Observations, Land Surface Models, and Ground Data: Implications for Data Assimilation. J. Hydrometeor., 5, 430–442.
Reigeluth CM, Schwartz E, 1989, An Instructional Theory for the Design of Computer-Based Simulations, Journal of Computer-Based Instruction, 16, 1-10.
Rochon GL, Loring NF, Jafvert CT, Stuart JA, Mohtar RH, Quansah J, and Martin A, 2006, Education in Sustainable Production in US Universities, Clean Technologies and Environmental Policy 8: 38-48.
Springer Verlag, 2004, Originally presented to the NATO CCMS Pilot Study on Clean Products & Processes Annual Meeting, May 2-6, Budapest, Hungary.
Rochon GL, Johannsen C, Landgrebe D, Engel B, Harbor J, Majumder S, and Biehl L, 2004, Remote Sensing for Monitoring Sustainability, In: Sikdar SK,

Glavič P, Jain R (eds) Technological Choices for Sustainability. Springer-Verlag Publishers, Berlin, Heidelberg & NY.

Rochon GL, Szlag D, Daniel FB, and Chifos C, 2002, Remote Sensing Applications for Sustainable Watershed Management and Food Security, In: Begni G (ed), Observing our Environment from Space: New Solutions for a New Millennium. A.A. Balkema Publishers, Lisse, The Netherlands.

Rogers DJ, Randolph S, 1993, Monitoring Trypanosomiasis in Space and Time, Parasitology 351: 739-741.

Sanderson SE, Redford KH, 2003, Contested Relationships between Biodiversity Conservation and Poverty Alleviation. Cambridge University Press, Cambridge.

Shepherd M., Burian S., 2003, Detection of Urban-Induced Rainfall Anomalies in a Major Coastal City. Earth Interactions, 7, 1–17.

Sims A., Niyogi D, Raman S, 2002, Adopting Drought Indices for Estimating Soil Moisture: A North Carolina case study, Geophysical Research Letters, 29 , 241 – 244.

Sinha AK, 2003, Development of an integrated disaster management system for India: Importance of reliable information. National Center for Disaster Management, India.

Tucker CJ, Choudhury BJ, 1997, Satellite remote sensing of drought conditions, Remote Sensing of Environment, 23, 243 – 251.

UNDP India, 2007, Disaster Preparedness and Response Plan, Ver. 45, 17 July 2007 version, Available from http://data.undp.org.in/dmweb/pp/UNDP_IDPR%2045%20Versionweb.pdf

USEPA, 2005, Technologies and techniques for early warning systems to monitor and evaluate drinking water quality: State-of-the-art review. U.S. EPA Office of Water, pp. 158.

Vajpeyi D, 2001, Deforestation, Environment and Sustainable Development: A Comparative Analysis. Praeger Press, Westport, Connecticut.

Vinodkumar, Chandrasekar A., Alapaty K., Niyogi D. 2008, The impact of assimilating soil moisture, surface temperature, and humidity and the traditional four dimensional data assimilation on the simulation of a monsoon depression over India using a mesoscale model, Journal of Applied Meteorology and Climatology, in press.

WCDR, 2005, World Conference on Disaster Reduction (WCDR), Kobe, Hyogo, Japan. January 18-22.

Wulder MA., Dymond CC., White JC, Leckie DG, Carroll AL, 2006: Surveying mountain pine beetle damage of forests: A review of remote sensing opportunities, Forest Ecology and Management, 221, 27-41.

Xavier VF., Chandrasekar A., Rahman H., Niyogi D., Alapaty K., 2008, The Effect of Assimilation of Satellite and Conventional Meteorological Data for the Prediction of a Monsoon Depression over India using a Mesoscale Model, Meteorology and Atmospheric Physics, in press.

6. Nearshore Coastal Processes Between Karwar and Bhatal, Central West Coast of India: Implications for Pollution Dispersion

Viswanath S. Hedge, G. Shalini, Shailesh Nayak and Ajay S. Rajawat

Abstract

In the present paper coastal processes between Karwar and Bhatkal, Central West Coast of India is discussed based on the Indian Remote Sensing program – 4 Ocean Color Monitor (IRS P4-OCM) data, wind and wave data and wave refraction derived near shore drift pattern, salinity condition and beach morphological changes. Near shore pollution dispersion pattern is inferred for the coastal region between Karwar and Bhatkal based on the observed coastal processes. Prevailing salinity conditions suggest upwelling during pre-monsoon to early monsoon, but become less effective during late monsoon, and down welling become important during winter. The upwelling are related to the combined effect of local wind and regional northward subsurface flow due to regional phenomena during pre-monsoon; South West-West approach of the waves for North -North West oriented coast and winter cooling effect during November and December respectively. The prevailing calm weather conditions due to low wave activities during December-January months facilitate tidal currents to dominate. The tidal currents move the water body towards seaward. Seasonal variations in the beach profiles indicated seaward movement of materials during monsoon, landward movement during post-monsoon and wave induced circulation patterns during pre-monsoon. In the vicinity of the river mouths, however, movement of the materials is dominantly northward. Sediments on the beach ranged from fine to medium grained, moderately to well sorted, which suggest dominance of waves on the coastal processes. Wave refraction derived alongshore drift pattern reveal seasonal variation in the direction of the current. IRS-P4 OCM data revealed 3 distinct pattern of sediment dispersion. Trend 1- during early monsoon, plume like sediment dispersion seaward especially in the mouths of the rivers; Trend 2- with the onset of monsoon, there is northerly transport of sediments with anticlockwise pattern with in the linear plumes; Trend-3 during post-monsoon, sediment concentration is less and it is southward. These patterns can be used to trace pollutants dispersion in the coastal sea.

6.1 Introduction

Rapid industrialization in the coastal belt, consequent waste disposal into the coastal water directly or via estuaries are causing the coastal sea pollution. Subsequent entry of the pollutants into food chain is causing severe health disorders. Pollution dispersion in the near shore water is highly complex and dependent on a number of factors such as the near shore coastal processes (including tidal currents, tidal range, wave currents, wave energy, wind forcing, wind direction, and hydrographic conditions) and geomorphic set up such as open coast, pocket beaches/bay type, and coastal orientation. In particular, complex coastal processes characterize the coastal regions of Uttara Kannada district, which exhibit a compound nature- submergence and emergence. Systematically gathered baseline information on coastal processes to address coastal pollution related issues are scarce for this region. A thermal power plant to generate electricity is proposed in the Tadari creek. In the present paper an attempt has been made to discuss the environmental implications of the project, and the near shore coastal pollution dispersion based on the ground truth data, wave refraction derived alongshore drift pattern.

In recent years, Ocean Colour Monitor (OCM) patterns and other satellite data products have been utilized successfully to trace total suspended sediments (TSS) dispersion (Rajawat 2005). Satellite images, in particular, Indian Remote Sensing (IRS) P4, having repitivity of alternate days and inbuilt sensors for oceanographic studies is used to understand sediment dynamics over regional scale and tracing the sediment transport pattern in space and time (Prasad *et al* 2002), suspended sediment concentration, sediment dynamics, circulation patterns and fronts, and consequent impact on coastal processes (Rajawat 2005). In view of the potentiality of OCM for application of coastal processes, in this study, IRS P4 OCM data have been used to compliment the ground based observations.

6.2 Methodology

Observations on the seasonal variations in hydrographic conditions were made along and across the near shore regions off Honnavar and Bhatkal from a fishing boat, and sampling locations have been recorded by a Garmin make Global Positioning System (GPS). Water samples collected in different seasons from the surface and near bottom samples using Niskin reversing bottle were studied for salinity, TSS and temperature following the standard procedure. The depth of sea bottom at sampling locations by fish weight method, time of water sampling and temperature of surface and subsurface water were recorded during the survey.

Seasonal variations in the foreshore morphology were monitored at selected beaches by stack and horizon method of Emery (1961). Simultaneously with beach profiling, foreshore sediments were collected for sedimentological study. Textural parameters were derived following Folk and Ward (1957) after preliminary treatments (Ingram 1970).

OCM data provided by National Remote Sensing Agency, India have eight bands both in BIL and BSQ mode. IRS-P4 OCM spectral channel, 412, 443, 490, 510, 555, 670, 780, 860 nm, with spatial resolution 360 m covering the coastal parts of the Karwar and Bhatkal region for different seasons were used to retrieve the suspended sediment concentration. Water leaving radiance was retrieved from OCM. The algorithm developed by Ramana *et al* (2000) for case II water is used for the present study. Suspended sediment concentration is derived using the relation

TSS = 25*Exp (2.16+ 0.991* log ((Rrs 555-Rrs 670) * (Rrs 555/Rrs 490))), where Rrs = Remote sensing reflectance.

6.3 Study area

The study area covers coastal and inner shelf regions between Karwar and Bhatkal along the Central West Coast of India (Fig. 1).

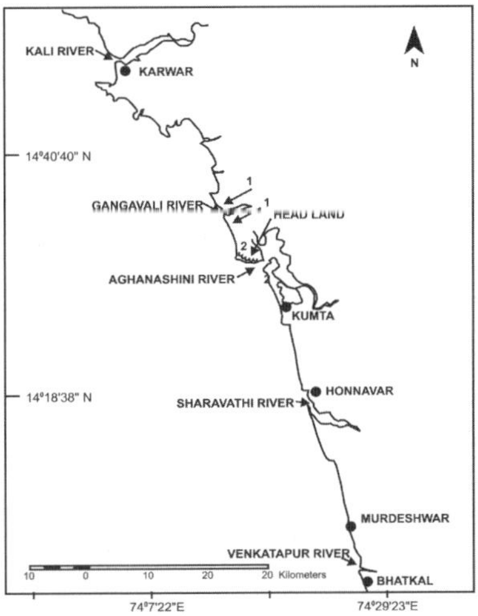

Fig. 1. Location map of the study area

The Kali, the Gangavali, the Aghanashini, and the Sharavati, are the major rivers that contribute sediment and water load into the Arabian Sea in this region. Due to construction of dam to the Sharavati and the Kali rivers, sediment budget in the down stream is affected. Due to growing urbanization, once a dense forest area, Western Ghats in this region is losing the vegetative cover. As a result of these anthropogenic activities, erosion of the basin and seaward transportation of both fine and coarse sediments have increased. All these activities are affecting the oceanographic process as well as sediment dispersal pattern in these regions. In addition to these activities, hydroelectric projects, Kaiga-Nuclear power stations, and many industries like manganese mining, Pulp and Ferro alloys in the Kali basin; Murdeshwar pilgrimage center and resulting anthropogenic activities, all contribute sediments and influence sediment dispersion as well as oceanographic process. In particular, Karwar Seabird Naval projects- related construction activities in the coastal region have caused major changes in the sediment dispersion in the region. Due to the completion of Konkan railway, highway facility, availability of hydro- power, the area is becoming a centre of developmental activities

6.4 Climate of the study area

The area experiences tropical climate, high rainfall around 300cm (June to September), and 85% of the rainfall during the South-West (SW) monsoon. Therefore, currents in the river mouth are controlled by fresh-water discharge during SW monsoon.

The tidal range is between 1- 3m and is referred to as meso tidal condition according to Short (1991). Wave energy conditions vary with season along the coast, being high during monsoon and low during fair weather season (Chandaramohan and Nayak 1991, 1994). Wave heights are less than 1 m during pre monsoon, 1 to 2m during post monsoon and 2 to 4 m during monsoon and define moderate wave energy regime especially during fair weather season (Davis 1984). Direction and velocity of wind also vary with season. During monsoon strong wind approach the coast from SW, where as during November and December they move seaward (Fig. 2).

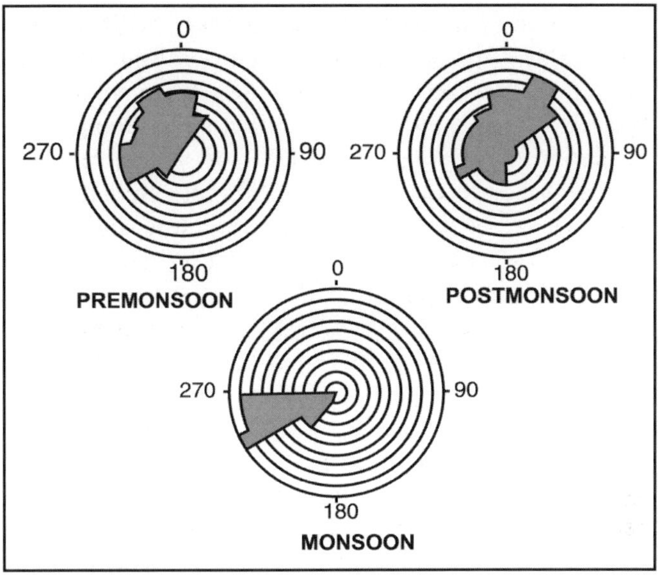

Fig. 2. Seasonal wind pattern in the study area

6.5 Geomorphology

The coastal region between Karwar and Bhatkal is complex in nature owing to the geomorphic features like spits (Fig. 3), head lands, cliffed shore line, lagoons, river mouth beaches in addition to the general North North West (NNW) orientation. Due to the spits growing across the river Gangavali from south to north, the river mouth is getting reduced and mouth is being shifted northward. As a result, rapid changes in estuarine morphology are taking place, which in turn is influencing the sedimentation processes. Presence of many islands in the inner shelf region off Karwar, Ankola, Kumta, Honnavar etc modify the wave refraction pattern.

Submerged -crescent shaped sand bars are being developed at the mouths of the rivers, which also influence sediment circulation by modifying the wave refraction pattern. Presence of a number of headlands along the coast obstructs the free sediment circulation across the beach (Fig. 4). These head lands are trapping sediments on one side favouring the beach development while on the other side provide shelter effect. Depending on the direction of wave approach, wave convergence and high energy at places, sediment starved beaches have become the site of erosion.

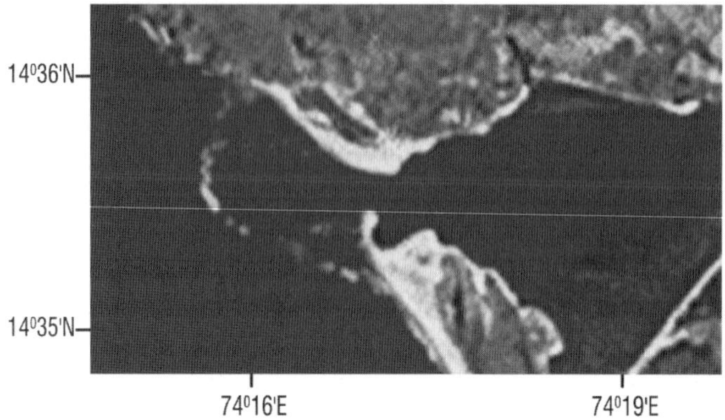

Fig. 3. Spit growth and submerged bar in front of the Gangavali river mouth near Ankola.

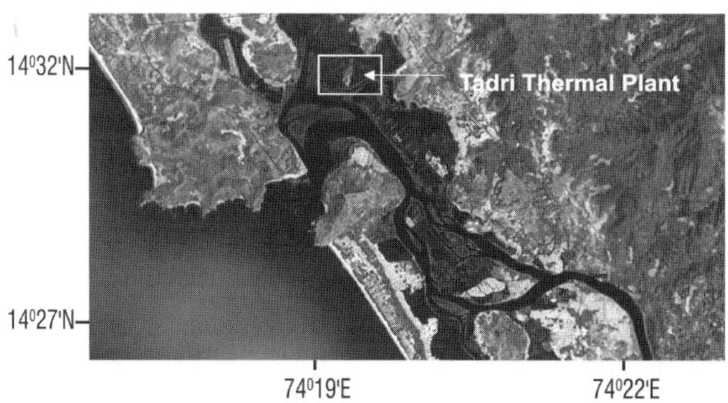

Fig. 4. Headlands on either side of the Aghanashini river mouth

6.6 Results and observations

6.6.1 Wave refraction analyses

During wave propagation from deep water to shallow water, waves undergo modifications in their height, speed and direction in response to bottom topography and shoaling. This process of wave transformation is called wave refraction and can be expressed as

$$\frac{\sin\theta_1}{C_1} = \frac{\sin\theta_2}{C_2} = \text{Constant}$$

where, θ_1= angle at the wave crest makes with the bottom contour over which it is passing

θ_2= a similar angle when it passes over the next contour

C_1 = wave velocity at the depth of first contour

C_2= wave velocity at the second contour

$C_1 = \sqrt{gD}$ in shallow water

$C = gT^2$ in deep water

D=depth of water, T=wave period, g=acceleration due to gravity.

The principle direction of wave approach in this region is from North West during November to February; from West during March to early June and from Southwest and South –Southwest during June to early November. The predominant wave periods range between 6 to 12 seconds. Therefore, wave refraction analysis is carried out for these wave conditions following Shore Protection Manual 1977.

Refraction pattern indicated that there is southerly alongshore drift for Northwesterly wave approach where as, for waves approaching from West, wave divergences occur in the river mouth region and wave convergence a little away from the mouths (Fig. 5). Presence of islands in front of the River Venkatapur, the Sharavati, the Kali etc may also aid this effect. These divergence and convergence of waves generate circulation cells in the surf zone (Murthy and Veerayya 1985). Southwesterly approaching waves observed during monsoon to early post monsoon generate northerly drift. As the energy of these waves is high compared to the post-monsoon, net northerly drift dominate over the southerly drift along the coast. Based on the geomorphic feature, Kunte and Wagle (2001) also infer net northerly drift in the surf zone.

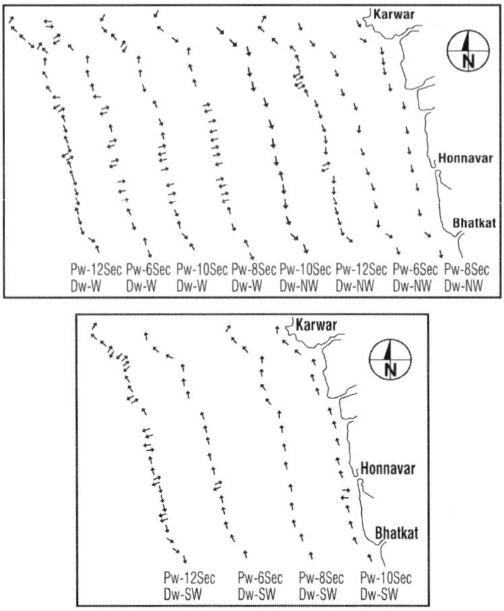

Fig. 5. Littoral drift pattern along the coast derived from wave refraction pattern for the study area

6.6.2 Salinity

Salinity in the ocean water ranges form 26.86 to 44.76 g/L in the study area. In general, the salinity profiles (Fig. 6) indicate distinct stratification. The regions adjacent to the mouth record low salinity (locations 1, 2,3,16, and 17) and those away from estuaries record higher salinity. Further, salinity during pre-monsoon is high as compared to the post monsoon, a feature common along the tropical coasts.

6.6.3 Temperature

Vertical gradient of temperature is less conspicuous (<2 C^0) in the study area. The increase in thermocline during summer is observed which is a common feature in the mid latitude (Pickard and Emery 1982) due to solar radiation. Temperature profiles for all the months show distinct difference between bottom, intermediate and surface waters.

6.6.4 Total suspended sediments (TSS)

During pre monsoon (March) concentration of TSS is high (> 30 mg/l) and low during October compared to the pre monsoon. During April there is indistinct difference between surface and bottom water with regard to TSS concentration. There are pockets of higher concentration in the near shore region (c.f.off Manki). In general, surface waters show higher concentration near the coast and show lesser concentration away from the coast.

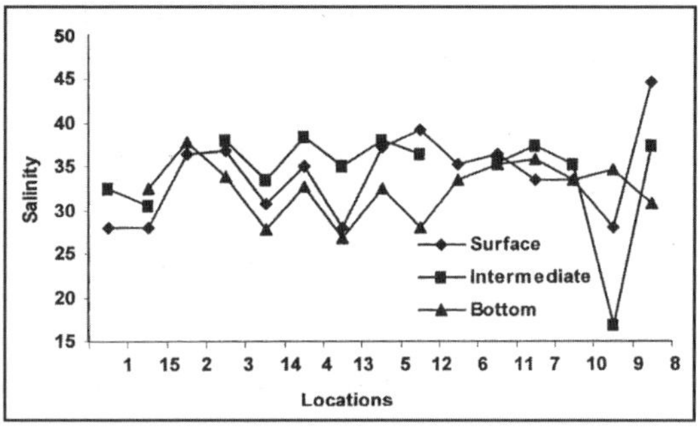

Fig. 6. Salinity profiles for the near shore region between Honnavar and Bhatkal

6.6.5 OCM pattern

Seasonal pattern of IRS P4 OCM for the different tidal conditions and seasons indicate distinct landward movement during pre monsoon (March) in the near-shore region, especially in the vicinity of the river mouths (Fig. 7a). Two different patterns of sediment transport are observed in the near-shore region. South of Bhatkal, sediment transport is southward, where as to the north of Bhatkal, they show many circulation patterns like local, landward and seaward.

FIG. 7 a

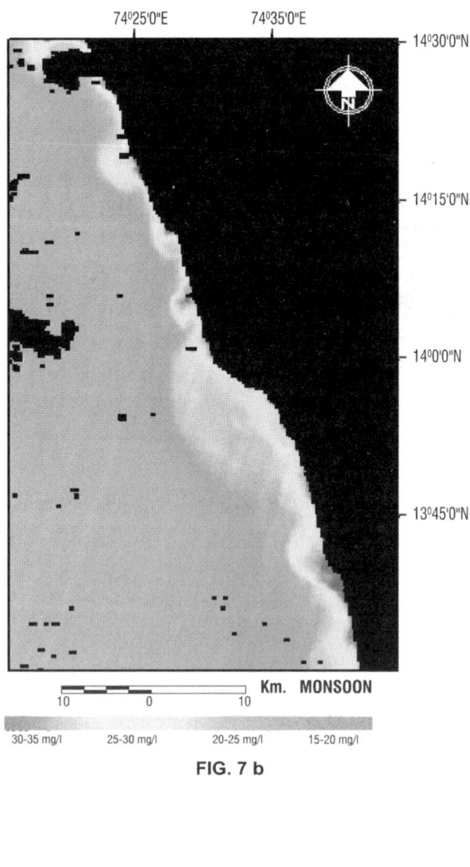
74°25'0"E
74°35'0"E
14°30'0"N
14°15'0"N
14°0'0"N
13°45'0"N
N
10
0
10
Km. MONSOON
30-35 mg/l
25-30 mg/l
20-25 mg/l
15-20 mg/l

FIG. 7 b

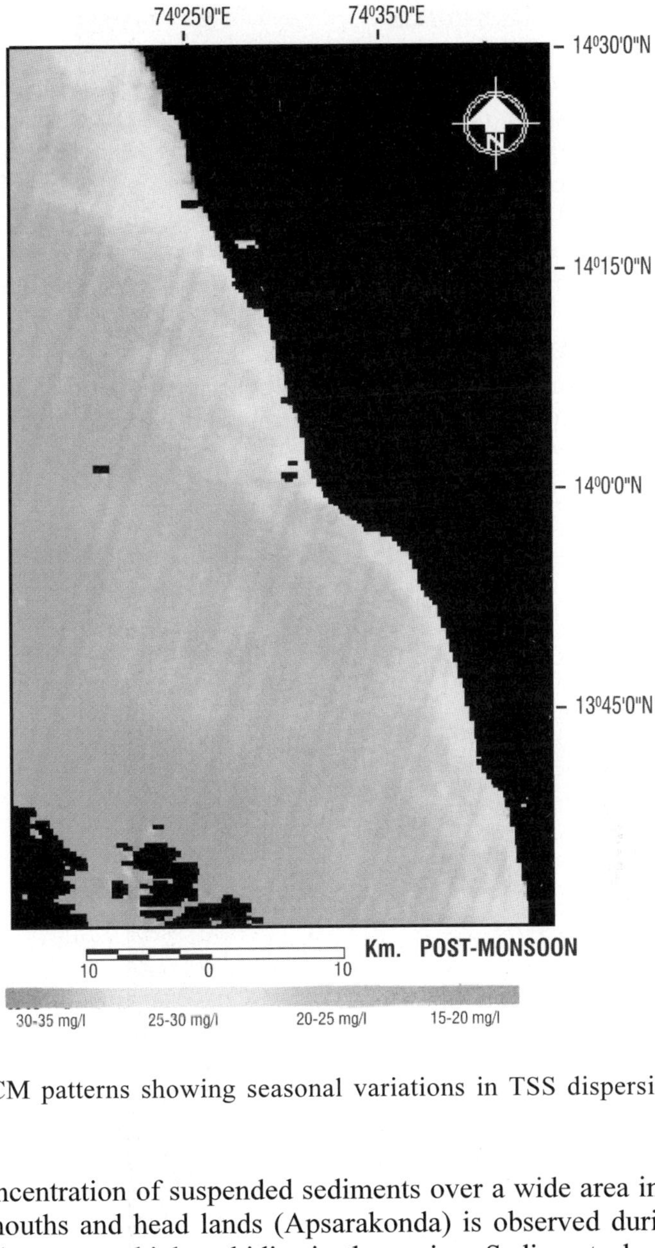

Fig. 7c. OCM patterns showing seasonal variations in TSS dispersion in the study area

High concentration of suspended sediments over a wide area in front of the river mouths and head lands (Apsarakonda) is observed during monsoon, which suggests high turbidity in the region. Sediment plume is observed in front of the Sharavati river mouth, the Venkatapur River, the Kali, the Aghanashini, and the Gangavali mouths (Fig. 7b). During post-monsoon (October), sediment concentration is less compared to pre monsoon and it is seaward (Fig. 7c).

6.6.6 Beach profile studies

Beach profile studies indicated accretion during post monsoon (September-October), erosion during monsoon (May-August), followed by a minor cycle of erosion during December / January (Fig. 8) which is a common feature of the beaches of the Central West Coast of India (Chavadi *et al* 1999; Bhat *et al* 2003). However, in the vicinity of the river mouths, it is highly variable especially during transitional periods (monsoon to post monsoon and pre-monsoon to monsoon). Similar depositional and erosional processes are observed along the beaches in the vicinity of the Kali River and the Gangavali river mouths (Nayak 1986; Chavadi *et al* 1989 respectively). During monsoon, erosion is observed both on either side of the river mouths, but during pre monsoon and post monsoon both the sides show different trends.

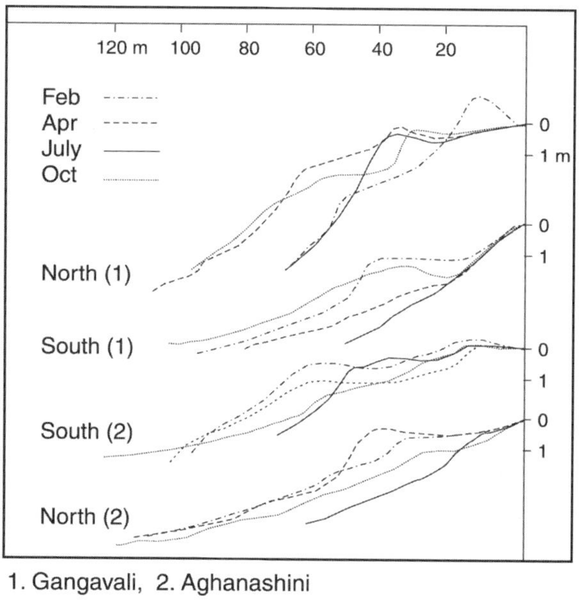

Fig. 8. Seasonal variation in beach profiles at selected locations in the study area.

6.6.7 Textural characters

Sediments along the coast are medium to fine grained (Table1), moderate to poorly sorted which suggest moderate to low wave energy conditions prevailing in the region. However, temporal and spatial variations in the grain size characteristics are evident. In general the sediments are bimodal to poly modal in nature and suggest either different sources or due to different mode of transport. The Coarsest first percentile versus median grain diameter (C-M) plots for the sediments of the beaches indicated uniform graded suspension and corroborate the above features. Coarsening of grain size close to the river mouths of the Gangavali (Chavadi and Hegde 1989; Bhat *et al* 2003); the Sharavati river mouth (Hegde *et al* 2007); the Kali river mouth (Hanamgond and Hegde 2002) during monsoon compared to pre monsoon have been reported which suggest selective removal of fines from the foreshore. Such a winnowing action results in improvement of sorting, but in the present case the sorting decreased. The bivarient plot indicates these could be due to river input.

Table 1 Textural characteristics of the beach sands in the study area

Location	Mean grain size (Phi)	Sorting SD
Majali Beach	2.68	0.41
Ramnath Beach	2.22	0.43
Sadashigad Beach	2.03	0.51
Kodibag Beach	1.71	0.47
Karwar Beach	2.23	0.48
Gangavali Beach	2.14	0.45
Aghanashini Beach	2.32	0.45
Kudle Beach	2.76	0.35
OM Beach	2.78	0.44
Kasarkod Beach	2.35	0.43
Manki Beach	1.94	0.47
Shirali Beach	1.95	0.48
Bhatkal Beach	1.41	0.36

6.7 Discussion

As the coast under study is oriented North-North West-South South East (NNW-SSE), there is always a component of wind stress parallel to the coast in an equatorward direction during early monsoon. This pushes the surface high saline water body southward favoring low saline bottom water to move upward leading to upwelling. However, salinity conditions do not suggest the indication of the Cold up welled water at surface during the months of September. With the onset of monsoon, wind stress is more northward owing to SSW approach of the waves. Due to high wave activity, freshwater discharged by the rivers is carried by the surface water currents and concentrated in the near shore. This floodwater may diffuse horizontally and vertically after it enters the sea and spreads offshore. It is also possible that large discharge by the river may prevent the cold-water upwelling off this coast. During winter, rivers coming from the western ghat bring cold water. In addition to these, low saline Bengal bay water meets northward flowing equatorial Indian Ocean water and flow northward as surface currents along the West Coast of India (*In* Harishkumar and Mathew 1997). This event causes reduction in the surface salinity along the SW coast. The net result is down welling in the region. The prevailing calm weather conditions due to low wave activities during November, December and January facilitate tidal currents to dominate. These tidal currents move the sediments seaward.

During February-March, the surface flow is southward due to wind approaching from NNW, Northwest and from West, where as subsurface flow is northward (*In* Harishkumar *et al* 1990). The northward flow undercurrent brings low saline water of equatorial origin towards north, and through the process of upwelling causes upward movement and reduces the surface salinity. When these water bodies move landward during rising tide, they reduce the salinity in the inner shelf.

Due to rough sea conditions during March, April and May, sea is agitated resulting in re-suspension of the silty materials. These conditions are conducive for silty- clay facies to move landward especially in the mouth regions. The submerged bar present in the mouth diffract the waves causing the materials to move alongshore.

Based on the model studies, Chandramohan and Nayak (1988) report southerly long shore current for the Central West Coast of India. However, the geomorphic features like spit growth towards north across rivers like the Sharavati, the Venkatapur, the Gangavali and the Kali suggest the dominance of northward transport especially in the vicinity of the river mouths. The observed TSS dispersion is land ward during high tide, especially during pre monsoon, and seaward during low tide. During post monsoon two distinct trends are observed both in terms of concentration as well as transport and during monsoon highly turbid conditions prevail. These features suggest a complicated sediment dispersal pattern along the coast.

6.8 Implications for pollution dispersion

It is well known that Indian coal is rich in Pb, Zn, Cr, Ni and SO_2 (Table 2). When coal is burnt, the concentration of these elements increases 6-7 fold in the ash produced (Table 3; c.f. Sahu 1987).

Table 2 Toxic heavy metals reported in coal (in ppm)

Elements	Indian Coal		U.S.A.Coal		Australian Coal		British Coal	
	Min	Max	Min	Max	Min	Max	Min	Max
Co	16	40	1	43	0.6	30	-	-
Cr	400	700	4	54	1.5	30	12	50
Ni	25	250	3	80	0.8	70	12	40
Zn	1000	2500	6	5350	15	500	-	-

(source: Sahu KC 1987)

Table 3 Toxic heavy metals in coal, coal-ash (in ppm) and its enrichment factors

Elements	Coal	Fly-ash	Enrichment Ash/Coal	Enrichments in soil 64 km away
As	0 5.9 ± 6.0	61 ± 6	10.3	1.8
Cd	0.19 ± 0.3	1.45	7.6	2.8
Cr	20.2 ± 5.0	131 ± 2	6.5	0.84
Co	6	38	6.3	1.9
Cu	18 ± 2 .0	120	6.7	1.4
Hg	0.12± 0.02	0.14± 0.01	0.9	0.5
Mn	40 ± 3.0	493 ± 7	12.3	0.8
Ni	15.1	98 ± 3	6.5	-
Pb	30 ± 9 .0	70 ± 4	2.3	2.2
Th	3.0	24	8.0	2.2
U	1.4	11.6 ±0.02	8.3	4.0
V	35 ± 3 .0	214 ± 8	6.1	0.9
Zn	37 ± 4 .0	210 ± 20	5.2	9,2

(source Sahu KC 1987)

Owing to high rainfall and SO_2 discharged into air by burnt coal, acid rain in the fragile ecosystem of the coastal region as well as on the Western Ghat occurs. The fly ash escaped from the smoke trap at the Chimney disperses onto the land during SW winds, and when drained by surface flow, they reach the sea. These may enter the food chain. In the sea, these pollutants get transported northward during monsoon and southward as well seaward during post-monsoon, thus causing entire coastal pollution. During pre-monsoonal upwelling (Hareeshkumar *et al* 1990) the deposited heavy metals get re-suspension and move landward during rising tide causing estuarine pollution. These waters are used during post-monsoon for aquaculture; the frawn culture is also affected. Due to winter down welling (Hareeshkumar and Mathew 1997) also there is re-suspension of the bottom sediments. At a number of locations, due to pocket beach or due to headland effects, pollutants are trapped on one side and get concentrated. Thus, the prevailing wind, tidal and current systems carry pollutants along the entire coastal regions of Uttara Kannada district.

6.9 Role of remote sensing

Satellite data due to synoptic view, multi-spectral and multi-temporal capabilities are extremely useful for monitoring coastal environment. Turbidity/Suspended sediments and chlorophyll concentrations are indicators of water quality that can be monitored using ocean colour sensors such as SeaWiFS, OCM, MODIS etc. mounted on satellites. Chlorophyll indicates trophic status, nutrient load and possibility of pollutants in coastal waters. Municipal sewage and industrial waste are major types of pollution observed on the coast. Such waste out-falls are difficult to detect as near shore waters are turbid. However, the information related to movement of suspended sediments helps in predicting waste effluent transportation path. In addition, ocean colour sensors are used to monitor phytoplankton blooms. Phytoplankton blooms are known to occur under various conditions in the near-shore regions. They may occur with cyclic regularity in certain regions where certain optimum environmental conditions prevail in marine waters. These planktons produce certain toxins, which adversely affect fish and other organisms. The bloom usually takes place rather suddenly and may spread with amazing speed, changing colour of surface water into red, green or hay colour. Although conventional point measurements of above discussed parameters using ships or boats provide accurate concentration measurements such surveys are limited to extremely poor spatial coverage that too of a particular time. Moreover, the costs of conducting surveys using ships/boats are very high. Efforts are in progress to simulate dispersal patterns of pollutants by integrating satellite based input parameters in hydrodynamic models.

Acknowledgements

We sincerely thank the financial support by the Indian Space Research Organization, Department of Space, and Government of India under ISRO-RESPOND program (10/4/359) and the Department of Ocean Development, Government of India (DOD/11-MRDF/1/4/P/98) in the form of research grants to carry out the present work. Authors sincerely thank Sri I.V.Ramana for the help is processing the OCM data in their laboratory. Authors sincerely thank the authorities of the SDM College of Engg. and Tech., Dharwad for the facility extended to carry out the work.

References

Bhat MS, Chavadi VC, Hegde VS (2003) Morphology and sediment movement in a monsoon influenced open beach at the Gangavali, near Gokarna (Central West Coast of India). Indian J Mar Sci 32(1): 31-36

Chandramohan P, Nayak BU (1991) Long shore sediment transport along the Indian coast. Indian J Mar Sci 20, pp.110-114.

Chandramohan P, Nayak BU, Raju VS (1988) Application of longshore transport equation to the Andhra coast, East Coast of India. J Coastal Engineering 12: 285-297 Chandramohan P, Sunil Kumar V, Nayak BU, Raju NSN (1994) Surf zone dynamics along the coast between Bhatkal and Ullal, West Coast of India. Indian J Mar Sci 23, pp.189-194.

Chavadi VC, Hegde VS (1989) A note on textural characteristics of the beach sands in the vicinity of the Gangavali river mouth, Uttara Kannada Dist, Karnataka. Mahasagar 22. pp.89-97.

Chavadi VC, Hegde VS, Nalawadi AR, Jalihal AA (1989) Erosion and deposition patterns of the beaches near Gangavali river mouth on the Northern Karnataka Coast. Geological Survey of India Spl Publ 24: 355-359.

Chavadi VC, Bannur CR, Hanamgond PT, Korkoppa MM (1999) Morphology and sediment characteristics of Uttara Kannada beach, West Coast of India. Indian J. Geomorphology 4 (1 &2): 9-34.

Davis JL (1964) A morphogenic approach to world shorelines. *Z.* Geomorphology 8: 27-42.

Davis JL, Hayes MO (1984) What is a wave-dominated coast? Mar Geol 80: 313-329.

Emery KO (1961) A simple method of measuring beach profiles. Limnol Oceanog 6: .90-93.

Folk RL, Ward WC (1957) Brazos river bar study on the significance of grain size parameters. J Sed Petrology 27: 514-529.

Hanamgond PT, Hegde VS (2002) Morphodynamic state of the beaches of Uttara Kannada district, West Coast of India, A monograph Pub SDMCET Dharwad: 75.

Hareeshkumar PV, Pradeepkumar T, Mathew B, Josheph MX (1990) Upper ocean thermal structure off Karwar (West Coast of India) during the withdrawal phase of monsson-1987, Indian J Mar Sci *19*(1): 36-41.

Hareeshkumar PV, Mathew B (1997) Salinity distribution in the Arabian Sea. Indian J Mar Sci 26: 271-277.

Hegde VS, Koti BK, Hanamgond PT, Shalini G, Girish KH (2007) Depositional environment of a tropical estuarine beach, located in the vicinity of Sharavati River mouth, Central West Coast of India. J Geol Soc India 69: 1279-1284.

Ingram RL (1970) Sieve analysis procedures, in: 'Sedimentary Petrology': 40-67. New-York, Wiley Inter-science.

Kunte PD, Wagle BG (1993) Determination of net Shore Drift Direction Of Central West Coast Of India Using Remotely Sensed Data. J Coast Res 9: 811-822.

Kunte PD, Wagle BG (2001) Littoral transport studies along West Coast of India-A review. Indian J Mar Sci 30: 57-64.

Murthy CS, Veerayya M (1985) Long shore current and associated sediment transport in the near shore areas of Kerala and Goa, West Coast of India. Mahasagar 18(2): 163-177.

Nayak GN (1996) Grain size parameters as indicator of sediment movement around a river mouth near Karwar, West Coast of India. Indian J Mar Sci 25: 346-348.

Nayak GN (1986) Studies of morphology, texture and mineralogy of the beaches along North Karnataka coast, around Karwar, India. PhD. thesis, Karnatak University, Dharwad: 210.

Pickard GL, Emery WJ (1982) Descriptive physical oceanography. Pergomon, Oxford: 249.

Ramana IV, Rao KH, Rao MV, Choudhury SB, Bhan SK (2000) Data processing scheme for the retrieval of oceanic parameters using IRS- P4 OCM data. Proceedings of the 5th PORSEC 2000, held at Goa, India during 5-8, December 2000: 765-769.

Prasad JS, Rajawat AS, Pradhan Y, Chouhan OS, Nayak SR (2002) Retrieval of sea velocities using sequential Ocean Colour Monitor (OCM) data. Indian Acad Sci. 3: 189-195.

Rajawat AS, Mukhesh Gupta, Pradhan Y, Thomoskutty AV, Nayak SR (2005) Coastal Processes along the Indian Coast- Case studies based on synergistic use of IRS-P4 OCM and IRS -1C/1D data. Indian J Mar Sci 34(4): 459-472.

Sahu KC (1987) Environmental impact of coal utilization in India-A geochemical approach. J Geol Soc India 30: 402-407.

Short AD (1991) Macro-Meso tidal beach morphodynamics- An overview. J Coast Res 72: 417-436

7. Landslide Hazard Zonation in Darjeeling Himalayas: a Case Study on Integration of IRS and SRTM Data

Mopur Surendranath, Saibal Ghosh, Timir B. Ghoshal
and Narayanaswamy Rajendran

Abstract

Landslide Hazard Zonation (LHZ) mapping of the Darjeeling Himalayas in Eastern India has been taken up by the Geological Survey of India to demarcate various zones based on their susceptibility to failure and suggest remedial measures. In this article, the various spatial data sets utilized and different digital techniques adopted in susceptibility mapping, the results of the GIS-based modeling studies following the BIS (1998) guidelines on macro-scale landslide zonation are described. The limitations of Digital Elevation of Models derived from topographic maps and IRS pan stereoscopic satellite imagery through digital photogrammetry are explained and the utility of 3-arc second Shuttle Radar Topography Mission (SRTM) DEM data to overcome those limitations in the highly rugged hilly terrains such as the Darjeeling Himalayas are discussed. Five zones in the order of increasing susceptibility are demarcated by GIS vector data overlaying techniques on the basis of Landslide Hazard Evaluation Factor (LHEF) rating scheme. The validity of this susceptibility zonation vis-à-vis existing landslide inventory is briefly dealt with and adoptability of BIS LHEF rating scheme while employing GIS-based modeling studies are critically evaluated.

7.1 Introduction

Landslides pose serious threat to human settlements, transportation, natural resources management and tourism in the Himalayan Region in India. Darjeeling Himalayas in West Bengal province are no exception, where every year during the monsoon period (between June and September), loss of human lives and colossal damage to properties take place due to slope instability. Darjeeling region witnessed devastating landslides in July 2003 resulting in the death of tens of people and complete disruption of communication network during the heavy rainfall. Geological Survey of India (GSI) has initiated on priority basis an integrated multidisciplinary programme on Landslide Hazard Zonation (LHZ) mapping of the entire Darjeeling region following the tragedy, to demarcate various zones based on their susceptibility to failure and suggest mitigative measures for minimizing the losses caused by the landslides.

7.1.1 Geological setup

The Darjeeling area in West Bengal, India is bound by Nepal Himalayas in the west, Sikkim Himalayas in the north, Bhutan Himalayas in the east and by the alluvial plains in the south. One of the most important factors contributing to the landsliding in study area is its complex geomorphological, geological and seismo-tectonic setup. The hill ranges of Darjeeling area are highly rugged, structurally-controlled and are constantly under the highly dynamic and active denudational (erosional) processes. Major Himalayan tectonic elements namely Main Boundary Thrust (MBT) and Main Central Thrust (MCT) traverse the southern parts of the area. The MCT separates the Proterozoic high grade rocks of Chutang Formation (present in the north) from the lower grade schists of Gorubathan Formation and Palaeozic Gondwana sediments. These rock formations are tectonically separated from the Siwalik sediments (Plio-Pleistocene age) by the MBT (Geological Map of India 1998). Several earthquakes ranging in magnitudes between 4.0 and 6.0 were reported in the recent past (Seismotectonic Atlas of India and its environs 2000).

7.1.2 Status of landslide hazard zonation studies

In India, landslide studies are conducted by a number of institutions, research and academic. GSI is one of the earliest to take up the landslide investigations in the Darjeeling area (Sinha et al. 1975). Since early 1980s GSI is undertaking landslide zonation studies in various hilly terrains of India. Besides, several workers of various organizations are involved in Landslide zonation studies in different parts of the major hilly terrains India. Saha and others (Saha et al. 2002) have attempted landslide hazard zonation in parts of Bhagirathi Valley using RS and raster GIS techniques. Ramakrishnan and others (Ramakrishnan et al. 2002) carried out LHZ in parts of Niligiri Hills while concentrating on the hill area development. Sudhakar and others (http://www.gisdevelopment.net / application / natural_hazards / landslides / nhls0006.htm) have carried out prioritization of landslide hazard zones in parts of Darjeeling district through Remote Sensing and GIS Approach. GIS-based statistical landslide susceptibility zonation - with a case study in the Himalayas was presented by Saha and others (Saha et al. 2005). The Building Materials and Technology Promotion Council (BMTPC), Ministry of Urban Development & Poverty Alleviation, New Delhi and Centre for Disaster Mitigation & Management (CDMM), Anna University, Chennai have prepared the landslide Hazard Zonation Atlas of India on 1:6,000,000 scale (2004) based on GIS techniques. Various departments of Indian Space Research Organisation (ISRO) namely NRSA, SAC, ADRIN and IIRS are also actively involved in the preparation of hazard zonation maps using RS and digital photogrammetry.

In view of a need for better coordination among various organizations and research groups so as to provide a focussed thrust to some critical aspects of landslide studies, the Department of Science and Technology (DST) has initiated a coordinated programme on the Study of Landslides which is being carried out in a multi-institutional mode. Geological Survey of India has been declared as the nodal agency for landslides by Government of India in 2004 (Ministry of Steel and Mines, communication no.11 [5] 04-M.I. dated 29.1.2004).

7.2 Methodology

Geological Survey of India undertakes various types of investigations on landslides - macro, micro and site-specific surveys and mapping mainly through conventional field methods. Bureau of Indian Standards (BIS) guidelines (BIS 1998) are being implemented by GSI in macro scale (1:25,000 / 1:50,000) landslide hazard zonation mapping in the various hilly terrains of India. BIS methodology, based on conventional field mapping, essentially comprises slope facet mapping of the terrain and subsequent grouping of these facets into various hazard zones based on the combined effect of several causative factors such as lithology, structure, slope morphometry, relative relief, land use and land cover, and hydrogeological conditions. In view of the developments in photogrammetry, particularly with the availability of high resolution remote sensing satellite imagery, GSI launched an integrated multidisciplinary programme on Landslide Hazard Zonation (LHZ) mapping of the entire Darjeeling region. Digital Photogrammetric techniques to derive Digital Elevation Models (DEMs) from satellite imagery, image processing tools for derivation of terrain parameters from DEM and GIS overlaying techniques to implement BIS guidelines in LHZ studies are employed for the first time by GSI.

Digital Elevation Modeling techniques are adopted to derive parameters on geomorphology-based landslide controlling factors like slope gradient, aspect, elevation, and plan and profile curvatures. Extraction of accurate DEM for the hilly terrain is not an easy task, keeping in view of the large area of present study (more than 1400 sq km) and the expensive nature of high resolution data and software techniques. A portion of the Darjeeling district has been selected to explore the optimality of various methods for the derivation of accurate DEM from different data sources such as topographic maps, IRS pan stereoscopic satellite imagery and freely available Shuttle Radar Topography Mission (SRTM) elevation data.

7.2.1 Digital photogrammetry studies on IRS PAN Data

Due to persistent cloud cover round the year only one pair of cloud-free IRS 1C pan stereoscopic satellite imagery (23.78 X 26.72 km. 5.8 m resolution, 13th November 1998 and 13th December 1998) with stereoscopic overlap of about 340 sq km area covering Darjeeling town and its environs could be acquired apart from one IRS pan satellite imagery (70 X 70 km. 5.8 m resolution, 9th November 2002). Erdas Imagine together with the Leica Photogrammetry Suite (LPS) is used for the extraction of DEM.

The raw IRS pan imageries after importing in Erdas Imagine were subjected to brightness inversion for effective usage in DEM extraction. During the course of DEM extraction from the stereoscopic data through LPS it was realized that Ground Control Points (GCPs) are essential to generate an accurate DEM. However, sufficient GCPs are not available for the study area. As an alternative, GCPs from the topo-maps (1:50,000 scale) are utilized. 2D root-mean-square error (RMSE) of about 0.6-pixel (less than 4 meters) was achieved during block bundle adjustment and Aerotriangulation. However 3D RMSE reported was more than 150 m in the DTM extraction stage in LPS OrthoBase Pro. In view of the minimum kernel size (3 X 3 pixels) during interpolation, a DEM with a resolution of 18 m was generated. The resultant DEM (Fig. 1) is in good quality, sharper and finer in details excepting a few holes and spikes.

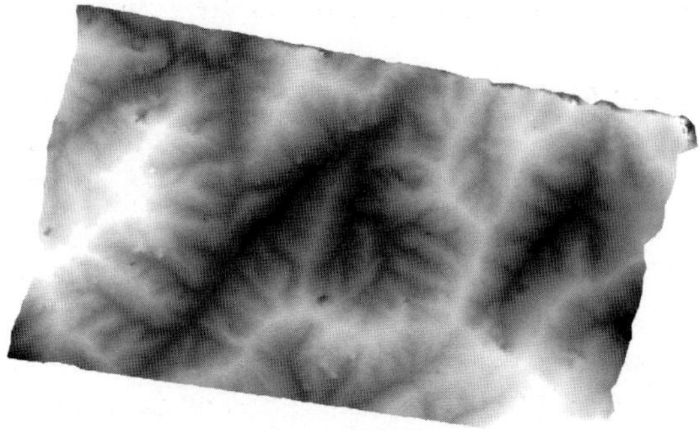

Fig. 1. DEM of Darjeeling generated from IRS 1C pan imagery

Due to the absence of GCPs and high 3D RMSE it was difficult to evaluate the resultant DEM for its elevation accuracy.

7.2.2 DEM Extraction from Topographic Maps

Earlier DEMs were mainly derived from the contour data (of different intervals) but digitization of contours of hilly and rugged terrains is laborious. Though it is difficult to extract elevation information for areas of deep gorges and escarpments, contour data has been utilized for the generation of DEM, since it is easily available from the topographic maps. Contours at 20 m and / or 40 m were extracted from the topo-maps of the study area and conventional interpolation techniques available with commercial software such as ArcGIS (with Spatial Analyst and 3D Analyst modules) were employed for extraction of DEM of 10 m resolution. Point data as elevation sources such as spot heights, triangulation points and bench marks were also extracted from the topo-maps. Break-line data such as drainage, escarpments and ridge crests are also included for refining the DEM.

Triangulated Irregular Networks (TIN) was derived from the contour data interpolation and subsequently these TINs were converted into DEMs of 10 m resolution. These DEMs appear to be of better quality in terms of vertical accuracy excepting the areas comprising deep gorges and steep valleys (Fig. 2).

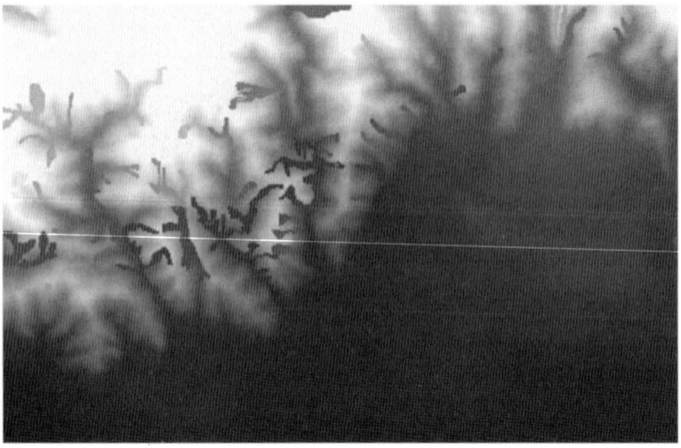

Fig. 2. DEM of a part of Darjeeling Himalayas based on contour data

7.2.3 Shuttle Radar Topography Mission (SRTM) DEM

In view of the limitations, of contour data interpolation techniques and digital photogrammetry on IRS 1C pan stereoscopic imagery, in deriving exact and accurate DEMs of this highly rugged study area, freely available SRTM elevation data was explored. In the present study, two SRTM one degree datasets (unfinished grade) N26E088.hgt and N27E088.hgt in compressed form covering the area between 88° and 89° E longitudes and 26° and 28° N latitudes in HGT format (16-bit signed integer raster data) were downloaded from Seamless Data Distribution System (ftp://edcsgs9.cr.usgs.gov/pub/data/srtm/) in December 2004. This dataset is a resample version of 1arc-second SRTM-1 with resolution of 3 arc-second (90 m).

The unedited SRTM data contain several voids or missing data patches due to lack of contrast in the original SRTM radar image, presence of water, or excessive atmospheric interference. These randomly distributed data holes or voids which are fewer in number in the study area, especially concentrated in steep regions (often on hillsides with a similar aspect due to shadowing) range in size from one pixel to regions of less than 2 km^2, have impeded the potential use of SRTM for any analysis.

Several spatial analysis techniques for filling the holes such as spatial filters, iterative hole filling and interpolation were available (Martin 2004) to improve the usability of SRTM DEM for terrain analysis. Void filling or patching of the missing data was performed by linear interpolation

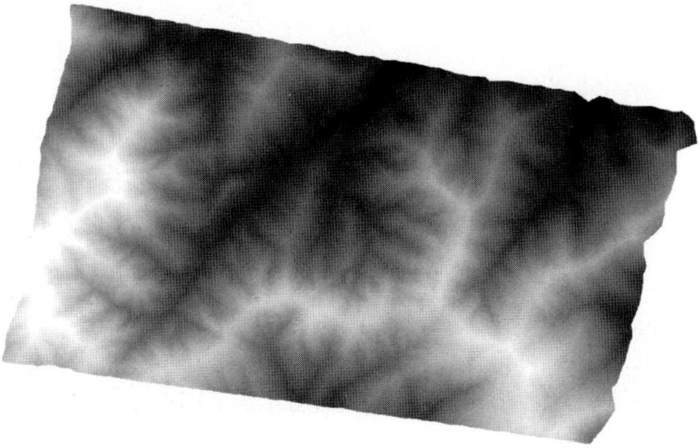

Fig. 3. SRTM DEM (corresponding to DEM of Photogrammetry)

algorithm available with the freely downloadable 3DEM software (http:// www.visualizationsoftware.com / 3dem.html). The resultant void-free DEM which is in Geographic projection on WGS 84 datum was reprojected on to Polyconic projection with Everest 1956 Spheroid and Indian (India-Bangladesh) Datum. The void-free DEM which is of 90 m resolution was resampled at 30 m resolution through nearest neighbour technique (Figs. 3 and 4) for deriving the terrain parameters. The extent of utility

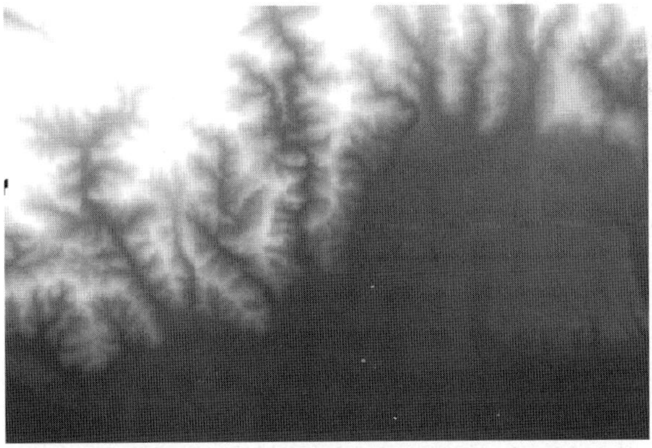

Fig. 4. Processed SRTM DEM (corresponding to DEM of contour data)

of DEMs extracted from the three sources described above, in deriving the terrain parameters pertaining the key landslide causative factors are described in Section 4.

7.3 GIS studies

7.3.1 Extraction of spatial datasets

The datasets required for Landslide Hazard Zonation (LHZ) as per BIS (1998) guidelines include lithology, structure, land use and land cover, slope morphometry, relative relief and hydrogeological conditions. The data pertaining to lithology and structure has been collected from existing geological maps (1:50,000) of the Darjeeling area. Digital image processing of IRS 1D satellite imagery (LISS 3 FCC and pan scenes) led to the extraction of land use / land cover and hydrogeology. Vector polygon datasets pertaining to these themes were generated in ArcGIS (ArcInfo).

Raster GIS tools available in ILWIS 3.3 Academic version, particularly filtering and reclassification tools were used to derive slope and slope aspect raster datasets from the DEMs created earlier. Five slope categories (<10°, 10°-25°, 25°-35°, 35°-45° and > 45°) and twelve slope aspect classes (each at 30° azimuth interval) were depicted in slope and slope aspect themes respectively. Assessment of combined effect of various landslide causative factors on each slope facet is the main focus in the present landslide macro-zonation studies. Therefore, slope facet maps (scale 1:50,000) were created by conversion of slope and slope aspect raster datasets into polygon themes and subsequent vector overlaying of these themes in ArcGIS (ArcInfo). Slope facets with less than 10,000 m2 area were included with the adjacent polygons in view of the scale of zonation (1:50,000).

Relative relief within each slope facet is calculated by raster zonal statistics in ArcGIS Spatial Analyst. Slope facet map as a zone dataset and DEM as a value raster were used to extract the minimum, maximum and range of elevations as statistical parameters within each facet. The elevation range field is renamed as relative relief for subsequent analysis.

7.3.2 Landslide hazard zonation mapping

Each of the aforesaid factor themes (spatial datasets) is assigned a rating (Table 1), a terrain susceptibility indicator to slope failure or slope instability, based on LHEF (Landslide Hazard Effective Factor) rating scheme (BIS 1998). The individual classes within each theme are rated subject to maximum rating of that particular theme. In case of slope factor theme there are five classes viz. 0-10°, 11°-25°, 26°-35°, 36°-45° and > 45° and the ratings are 0.5, 0.8, 1.2, 1.7 and 2.0, respectively. Similarly relative relief

Table 1. LHEF rating scheme

Theme	Maximum LHEF Rating
Lithology	2
Structure	2
Slope Morphometry	2
Relative Relief	1
Land use and Land cover	2
Hydrogeological conditions	1

Table 2. Various hazard zones based on TEHD Values

Zone	TEHD Value	Description of the Zone
I	<3.5	VLH (very low hazard) zone
II	3.5 to 5	LH (low hazard) zone
III	5.1 to 6	MH (moderate hazard) zone
IV	6.1 to 7.5	HH (high hazard) zone
V	>7.5	VHH (very high hazard) zone

has three classes viz., < 100 m, 100-300 m and >300 m and the ratings are 0.3, 0.6 and 1.0, respectively. Every individual feature (facet) in the Facet map is tagged with LHEF rating attribute for different factor themes. Total Estimated Hazard (TEHD) on each facet is calculated by combining the cumulative effective rating (a 10 point-grading) of all the six factor themes. Five zones, from very low hazard to very high hazard, have been identified based on TEHD value ranges (Table 2). Landslide Hazard Zonation has been carried out by assigning each facet to a particular hazard zone using the TEHD value and a zonation map for the south-eastern part of Darjeeling Himalayas (Fig. 5) was generated.

Table 3. Various hazard zones affected by landslides

Susceptible zone (Sq. m.)	% of zone in total area	% of Zone affected by landslides
VLH (very low hazard) zone	1.63	0.56%
LH (low hazard) zone	56.7	19.52%
MH (moderate hazard) zone	108.0	37.24%
HH (high hazard) zone	113.0	39.01%

7.4 Discussion

7.4.1 Validation of LHZ mapping

In order to validate the zonation mapping, the existing landslide inventory theme is superimposed over the hazard zonation map. More than 75% of the area (Table 3) falls under moderate (37.24%) and high hazard (39.01%) zones whereas about 4% is within the Very High Susceptible Zone. It further showed that 58% of the area in high susceptible zone, 28.37% in moderate susceptible zone and 5.8% in Very High Susceptible Zone affected by landslides. The landslide inventory further showed that concentration of landslide activity mainly in the southern part of the area though the terrain in the northern part comprises several HH and MH zones. This anomaly in landslide activity can be attributed to the presence of Main Central Thrust (MCT) and Main Boundary Fault (MBT) in the southern part. Another reason could be due to the presence of soft younger sedimentary rocks in the south, whereas the northern part is represented by the high grade metamorphic rocks. Therefore the hazard zonation map in this study need to be refined further by taking some more parameters into consideration particularly in areas which are tectonically and seismically active. It may also be required to modify the weights for different parameters and classes within each parameter.

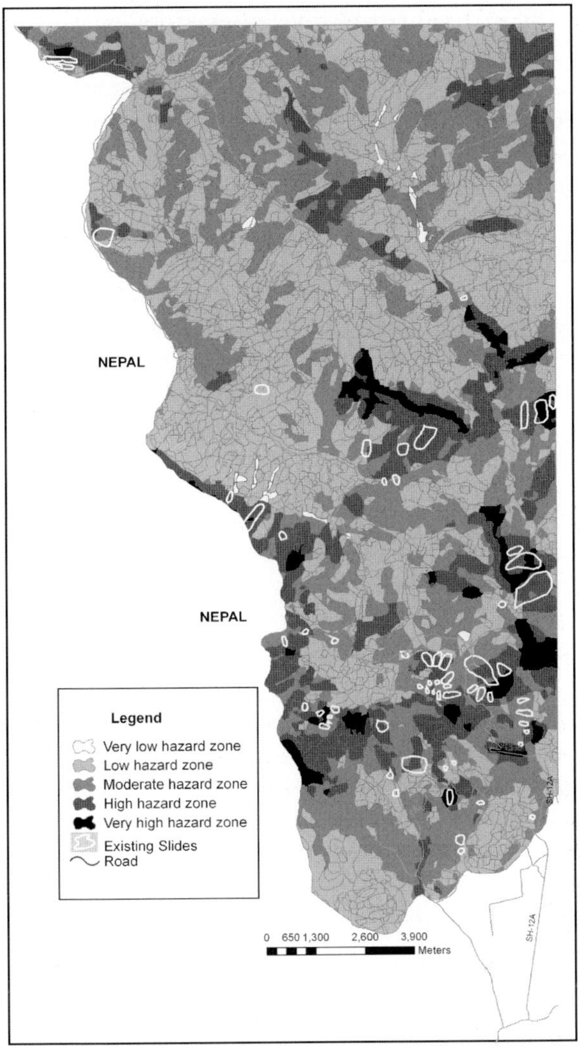

Fig. 5. LHZ map of part of Darjeeling Himalayas

7.4.2 Limitations of BIS techniques in LHZ

The preparation of landslide zonation or susceptibility maps are of great importance to planning agencies for preliminary hazard studies especially when a regulatory planning policy is to be implemented. Small-scale regional surveys are low cost techniques by which large areas can be covered in a relatively short time permitting an economical and rapid hazard assessment. To this extent BIS technique of hazard zonation over the base facet maps may be of considerable help. With the rapid urbanization and spread of developmental activities in the hilly terrains, more and more areas are being affected by landslides and therefore one has to consider other causative factors as well particularly, the anthropogenic interference

Apart from the six causative factors, vegetation, pedology and climate (particularly rainfall) are also among the most inducive factors and to be taken up exclusively instead of combining with other factors.

The concept of slope facet as the basic study unit needs to be re-examined. Excepting the isotropism in case of slope morphometry no other causative factor can have uniform properties, whether it is lithology, structure, land use or hydrogeology. The relationship between slope of the each and every facet and geological structures such as joints, fractures and bedding planes in terms of inclination and direction, required to be established. In the terrains such as the present area which is highly rugged and covered it may not be possible to assign LHEF rating for this parameter in all the facets. Modern and proven landslide hazard assessment techniques such as landslide inventory analysis, heuristic analysis, statistical analysis and deterministic analysis (van Westen 1994) should be applied instead of this technique which had been devised mainly for conventional field / manual surveys.

7.4.3 Optimality of DEM techniques

DEM studies are a must for landslide hazard zonation since it is the only digital source for extraction of parameters on slope-controlled causative factors. Extraction of DEM from conventional data sources such as topo-maps or contour maps is one of the oldest techniques which is still followed in developing countries like India as they are easily available. Due to the presence of numerous deep and narrow valleys or gorges, escarpments and cliffs in the Darjeeling Himalayas, contour data is incomplete. Hence the terrain parameters derived from these DEMs can not be relied upon especially in areas having narrow valleys and escarpments.

Though the required GCPS are absent, a moderate quality DEM of 18 metre resolution, for a small portion of the study area was extracted from the IRS pan stereoscopic satellite imagery. The high cost of stereoscopic satellite data and software and the non-availability of accurate GCPs prevent from adopting the photogrammetric method for DEM extraction in small scale or regional landslide hazard zonation studies of a large region like Darjeeling Himalayas.

3 arc-second SRTM DEMs which are processed and resampled to generate void-free 30 metre resolution DEMs are thoroughly explored and found to be good for rapid generation of slope morphometry maps on 1:50000 scale for large areas like the present study area. Missing patches are not many in the SRTM DEM of this region and they can be easily eliminated by linear interpolation techniques and hence are not at all a constraint for this study. Jarvis (Jarvis et al. 2004) who gave a good comparison of SRTM DEM with DEMs generated from cartographic data, concluded that SRTM DEM are better for deriving slope morphometry parameters rather than DEMs generated from 1:50,000 topo-maps. Processing of SRTM DEM may not be a hindrance with the availability of processed SRTM DEMs (5 X 5 degree tiles of 90 m resolution) freely from Consortium for Spatial Information (CGIAR-CSI) and can be used straight in regional landslide hazard zonation studies.

7.4.4 Vector GIS techniques in landslide hazard zonation

Vector polygonal factor themes were generated for various causative factors of slope instability as per BIS standards. In view of the macro zonation studies carried out on 1:50,000 scale, many facets (polygons) were included with adjacent classes in facet map, so as to keep the minimum size of the facet to 10,000 metre^2, which resulted in having facets with erroneous slope morphometry values. In stead of overlaying factor maps which is the usual practice in GIS analysis of landslide zonation, in the present case effective factor ratings for each theme are calculated on each facet. While doing so the polygonal area of the individual classes within each factor theme versus the size of the facet determines the class of the factor which will be assigned to the facet. As a result the actual effect of various causative factors of slope instability is missing partially or completely.

7.5 Conclusions

The present study clearly indicates the advantage of the processed 3 arc-second SRTM DEM in deriving terrain parameters for regional landslide zonation studies (1:50,000 scale) and the limitations of topo-maps in DEM generation pertaining to hilly regions. If good and accurate GCPs are available then one can choose digital photogrammetry of high resolution stereoscopic imagery as well for DEM generation provided they are available. The use of BIS techniques in LHZ may be critically re-examined while adopting digital techniques for hazard zonation. Instead, modern and proven assessment techniques such as landslide inventory analysis, heuristic analysis, statistical analysis and deterministic analysis through GIS (raster analysis) should be applied.

References

Andy Jarvis, Jorge Rubiano, Andy Nelson, Andrew Farrow, Mark Mulligan (2004) Practical use of SRTM data in the tropics: Comparisons with digital elevation models generated from cartographic data, International Centre for Tropical Agriculture (CIAT) and the International Plant Genetic Resources Institute (IPGRI), Cali, Colombia, Working Document no. 198

BIS (1998) Indian Standard Preparation of landslide hazard zonation maps in mountainous terrains – guidelines, Part 2 macro-zonation, Bureau of Indian Standards, New Delhi, India, IS:44496 (Part2)

Consortium for Spatial Information (CGIAR-CSI) of the Consultative Group for International Agricultural Research (CGIAR) (2005) SRTM Data Processing Methodology, http://srtm.csi.cgiar.org/SRTMdataProcessingMethodology.asp Accessed 27 May 2006

Foumelis M, Lekkas E, Parcharidis I (2004) Space Data from Earth Sciences. Elsevier, Amsterdam, pp 321-332

Geological Map of India (1998) Geological Survey of India, Kolkata, India.

Hengl T, Gruber S, Shreshta DP (2003) Digital Terrain Analysis in ILWIS, Lecture notes and user guide. International Institute for Geo-Information Science and Earth Observation (ITC), Enschede, Netherlands

Landslide Hazard Zonation Atlas of India (2004) Building Material and Technology Promotion Council, New Delhi, India

Martin G (2004) Free and low cost datasets for international mountain cartography. http://www.icc.es/ workshop/ abstracts/ica_paper_web3.pdf. Accessed 24 May 2006

Ramakrishnan S, Santee Kumar V, Afar Sadie MGSM, Venugopal K (2002) Landslide zonation for hill area development http://www.gisdevelopment.net/application/natural_hazards/landslides /nhls0013.htm. Accessed 27 May 2006

Saha A, Gupta RP, Arora MK (2002) GIS-based landslide hazard zonation in the Bhagirathi (Ganga) valley, Himalayas. Int J Remote Sens 23(2):357-369

Saha AK, Gupta RP, Sarkar I, Arora MK, Csaplovics E (2005) An approach for GIS-based statistical landslide susceptibility zonation - with a case study in the Himalayas. Landslides 2(1):61-69

Seismotectonic Atlas of India and its environs (2000) Geological Survey of India, Kolkata, India, pp 12-13

Sinha BN, Verma RS, Paul DK (1975) Landslides in Darjeeling District, West Bengal and adjacent areas, Bulletin of the Geological Survey of India Series B, Engineering Geology and Groundwater, no. 36

Sudhakar S, Sadasivarao B, Suryaprakasa Rao B, Venkateswara Rao V Prioritization of landslide hazard zones in parts of Darjeeling district, West Bengal, India- a RS and GIS approach. http://www.gisdevelopment.net/application/nrm/mountain/mounm0004.htm. Accessed 27 May 2006

Surendranath M, Rajendran, N (2005) Digital terrain analysis in landslide hazard zonation: A case study, Erdas India User Conference, Hyderabad, India, January 3, 2005

Van Westen CJ (1994) GIS in landslide hazard zonation: a review, with examples from the Colombian Andes. In: Price MF and Heywood DI (eds) Mountain Environments and Geographic Information Systems. Taylor & Francis, London, pp 135-166

8. Monitoring and Interpretation of Urban Land Subsidence Using Radar Interferometric Time Series and Multi-Source GIS Database

Swati Gehlot and Ramon F. Hanssen

Abstract

Subtle and gradual land subsidence in the Netherlands is a well-known phenomenon and it potentially involves a high risk of subsidence disaster in the densely populated areas. Based on the conventional geodetic techniques such as leveling, the slow deformations in urban areas are difficult to monitor due to low temporal and spatial sampling frequencies. We present the potential of Persistent Scatterer Interferometry (PSI) as an efficient satellite based technique of deformation monitoring. Time series of radar multi-epoch images (up-to 15 years) are analyzed to extract the targets with a stable phase behavior over time. PSI is particularly useful in application to urban environment as the buildings and solid structures are favorable radar point targets.

For the interpretation of the PS data, we propose an approach of multi-source geo-information combination. Geographical Information Sys-tem is used to combine different geo-information sources such as land-use, cadaster databases, aerial photos and digital ortho-map. This combination leads to the possibility of interpretation of PS data with reference to physical ground features, and hence, provides an idea of the reasons of the subsidence. Moreover, a Web-GIS interactive visualization interface is designed to share the PS derived land subsidence data to a variety of users. Test cases of PSI analysis for the cities of Delft and Rotterdam in the Netherlands are presented and discussed.

8.1 Introduction

Country like the Netherlands, with most of its areas having ground level below mean sea level, is highly susceptible for land subsidence and flooding risk. The conventional geodetic techniques for measuring the ground surface changes such as levelling etc. are carried out on a countrywide scale. However, the spatial data sampling density in these conventional deformation monitoring methods is highly insufficient compared to what is needed to explore and interpret the slow deformations. These subtle and slow localized deformations arise due to various driving mechanisms, for example, infrastructural works, industrial water extraction, ground water and soil variability, faulty building foundations, etc. Moreover, the conventional field based techniques are time consuming as well as very expensive to carry out, resulting in the low temporal sampling of the data. Many levelling campaigns take as long as 4 to 5 years to finish and as a result, many small or localized areas undergoing subsidence remain unnoticed. Also, this subsidence of ground with reference to the sea level leads to a gradual accumulation of a high flooding risk in the areas close to the dikes along the coastline.

In this paper we present the ongoing research in land subsidence monitoring carried out at the Delft University of Technology. A satellite based ground deformation monitoring technique under development at the Delft Institute of Earth Observation and Space Systems (DEOS) is discussed in this paper. This technique, called as the Persistent Scatterer Interferometric Synthetic Aperture Radar (PS-InSAR), uses multi-epoch radar data to compute the average displacement rate of radar targets in the area under study. The reflection characteristics of the radar signal are analyzed in the data time-series to extract the pixels with time-coherent phase behavior. These selected pixels are referred to as Persistent Scatterers or PS. This paper provides an overview of the PSI technique and results of some case studies carried out in the Netherlands. We present the PSI as a potential deformation monitoring technique, giving examples of results and interpretation in the densely populated areas.

This research stresses the need of incorporating supplementary geo- information sources as an approach for improved interpretation of the deformation. After combination of various data, a MapServer WebGIS tool is used for dissemination of PS data to the user community (for example, decision making authorities, planners etc.) using a simple web browser. In section 2, an overview on the processing steps of the PSI is described. Section 3 and 4 discuss the main problems in the interpretation of PS observations and the web-GIS approach of PS visualization respectively. Test cases for Rotterdam and Delft cities in the Netherlands are shown in section 5. Finally, the conclusions are derived along with a note on further direction of work.

8.2 Overview: Persistent Scatterer Interferometry (PSI)

As a supplement to various conventional subsidence monitoring techniques, InSAR (Interferometric Synthetic Aperture Radar) has been used since the last two decades. InSAR gives the possibility to measure the deformations of the earth's surface to sub-cm level of precision. Numerous studies have been reported about the land subsidence/deformation consequent to earthquakes, hydrocarbon extraction, ice motion, changes in the ground water flow and geological phenomena (Massonnet, and Feigl (1998)).

However, conventional InSAR applications suffer the inherent problems of geometric and temporal decorrelation of the deformation signal signal, along with the atmospheric variability over space and time (Hanssen, 2001). The geometrical decorrelation corresponds to the measurement noise due to the different look angles of the two satellite passes, and temporal de-correlation arises as a result of change in scattering characteristics of the ground objects with time.

Persistent Scatterer Interferometry (PSI) (Ferretti et al., 2001) offers an advantage over the conventional interferometry to handle the geometric decorrelation, temporal decorrelation and the atmospheric artefacts. PSI exploits a temporal data archive of radar satellite images acquired over the deforming region. With satellites ERS -1 and -2 followed by ENVISAT launched in 2002, large amount of radar image data sets are available for analysis. Data archives of time periods up to 15 years (or 30 to 100 images) are useful in analyzing the behavior of subtle and slow deformation of radar point targets.

Persistent Scatterers are time-coherent pixels identified from a stack of differential interferograms that have a common master image. In principal, the PSs are the pixels on the radar image that show similar reflection characteristics over a long period of time. A master image is chosen from the available SAR images on the basis of a favourable geometry in relation to other images, i.e., optimal relative temporal and perpendicular baseline, and Doppler centroid. After the co-registration and oversampling (2 times) of the master and the slave images, a series of interferograms is created with the use of precise orbit information. In the current work, DORIS (Delft Object Oriented Radar Interferometric Software) is used for these processing steps. Various combinations of radar data time series have been used in the case studies discussed here.

The subsequent processing for the detection of Persistent Scatterers is done using the Delft PSI algorithm. Coherent targets are identified by their amplitude and phase behavior before and after estimating topographic, displacement, and atmospheric phase contributions. Time coherence estimation is performed for individual PS and serves as a quality parameter for the measurements. The number of selected PS locations depends on the coherence threshold value — the higher the coherence threshold, the lower the number of accepted PS points. At the end of the PSI processing chain, the database of estimated parameters consists of locations *(X,Y)*, relative topographic heights, displacement rates, time coherence, displacement time series, and the atmospheric signal time series for an individual PS (Gehlot et al., 2005).

Persistent Scatterers usually correspond to solid or man-made structures, which commonly have a high radar backscatter (for example, buildings, lampposts, street/road edges, exposed rocks, solid surfaces etc.). The number of detected PS is therefore higher in an urban area in comparison to a rural or vegetated area. Moreover, the Persistent Scatterers or the deformation profile locations are usually much more in number compared to those derived by levelling and other conventional techniques of deformation monitoring. This shows an advantage of detection of so many deformation profiles from satellite images, without the need of fieldwork or expensive ground survey. This gives an idea about the cost-effectiveness of the PS-InSAR technique in comparison to other traditional methods.

8.3 Interpretation of PSI results and the use of web-based Geographical Information System

Although the PS observations are very precise, the interpretation of the PS observations is rather challenging. This is due to the fact that the complex radar signal is difficult to reconstruct in terms of scattering mechanism of features on ground. Hence, to decompose the signal into the parameter of interest, i.e., displacement needs supplementary information sources about the ground objects (or the radar targets). Relating the PS behavior to local geo-information is an important step to interpret the signal, and further, correlating it to the mechanism(s) related to the observed subsidence behavior.

In this study, a Geographical Information System (GIS) is included as an interactive tool to visualize and interpret the radar time series data. Incorporation of supplementary geo-information pertaining to physical features in the area gives a better understanding of the derived deformation by PSI. Useful information sources are cadastre databases, land-use maps, vector data sets of road and house construction boundaries and shallow and deep geology. The present research shows a combination of the PS technique with a dedicated GIS toolbox to assist the PS data interpretation. Various case studies performed in different regions of the Netherlands revealed substantial deformation rates of couple of millimetres per year.The following are the commonly available and used geo-data sources in context of the Netherlands.

8.3.1 Raster Data

1. The Dutch topographic map (scale 1:25 000) used as a backdrop raster image for a visual inspection of the PS locations with respect to various man-made features.
2. A high-resolution (1 meter) aerial orthophoto map of the area is also used for a quick reference to visualize PS.
3. The radar Multi-image Reflectivity Map (MRM) is used after geocoing the image from radar coordinates to Dutch RD coordinates. The MRM is computed by averaging the radar images of the study area. Further, this averaged image is multi-looked by a factor of 5 in azimuth to account for the different resolutions in range and azimuth direction.

8.4 Vector Data

The *Top10* Vector Map of the Netherlands, maintained by the Land Registry Office at Netherlands (Top10 Vector, 2006) is used in classification of land use. The map scale is 1:10 000 and it completely covers the features of railroads, highways, infrastructural boundaries of underground metro tunnels, hydrographical features, houses, streets, built-up areas, and other such land use classes. The *Top10* vector map is divided into different land use classes as line and polygon vector dataset. The following list covers the specific land-use information data used in this work,

- Line vector for boundaries of infrastructures, namely railroads, transmission lines, metro and tram lines etc.
- Polygon vector for boundaries of water areas.
- Polygon vector for boundaries of forests and vegetated areas.
- Polygon vector for buildings and houses areas.
- Line vector for boundaries of buildings and built-up areas. This map is called as *large Scale Standard Map* and abbreviated to GBKN (GBKN, 2006) in Dutch. In suburban areas, the scale of this map is 1:500 or 1:1000.

8.5 Web-GIS Visualization of PS Data

Ground deformation data like the urban land subsidence or behavior of infrastructures, bridges, dykes etc. are most useful for the organizations like the city authorities, municipal departments, planners, infrastructure industries, mine industries, and Non-Government Organizations (NGOs). However, due to the complexities in the PS-InSAR data processing, the technique is often too difficult for this "non-expert" user community. Moreover, these parties are interested in the final results with reference to the existing information. For example, if city authorities are able to visualize the PS results on the backdrop layer of infrastructural foundation details, they can quickly act on this information. Moreover, end users, with specific expertise can aid in better characterization of the radar scatterers, aiding further advancement in the technique. On the other hand, the providers of the PS results, i.e., the PSI scientists are not well aware of the local areas so as to make a correct and localized interpretation of land subsidence detected by PSI.

To bridge this gap between the providers of the PSI technique i.e., the scientists and the end user community, we explore the possibility of using Web-GIS for an interactive visualization of PSI results. A Mapserver Web-GIS tool (Mapserver, 2006) is used to create the Internet visualization interface of PS-InSAR data. Originally developed by the University of Minnesota, Mapserver is one of the CGIs (Common Gateway Interface) that allow a server to geographically present/display the data in form of a "map". Since Mapserver is designed and developed to support the evolving OGC standards (Open Geospatial Consortium OGC, 2005), it can be used to incorporate any remote data sources that publish data consistent with these standards. One of the main advantages of a web-based system is the ability to provide mapping capability to the general community i.e. anyone who can run a Web-browser, even in low bandwidth conditions, can visualize the PS results. Gigabytes of spatial data can be manipulated on the server side and only small compressed images (usually less than 100 KB) are sent to the client. Via Mapserver, the dedicated GIS operations on the server can be called with simple buttons on the client or the internet explorer. The HTML (Hyper Text Markup Language) client or the browser with a query string sends variable values to the server, and the Mapserver CGI parses/renders the variables, reads the digital geo-data (in form of shapefile (ESRI, 1998)) described in the Mapfile, draws the requested map, prepares an HTML page (according to template definition) for publishing and sends this HTML page to the client browser. This process continues each time the browser sends in a variable request to the server.

A Mapfile (UMN Mapserver, 2006) or simply an input file is created for the Mapserver to connect the stored data shapefiles (ESRI, 1998) and database of the persistent scatterers and the supplementary information data. This Mapfile is an ASCII file that lists the details of the input and output data attributes, map bounding box, etc. The attribute appearances, colour scales, and definition symbology for the desired display of vector data (points, lines, polygons) and raster data can be controlled by the layer definition in the mapfile. The Mapserver CGI is called by the means of an HTML template linked to the Mapfile.

The visualization interface is initialized by typing in URL (Mapserver link url to the data server of PS + the geo-data) to the browser address bar. PS data points as vector objects are overlaid with the other datasets to have a better understanding of the PS behavior according to ground features. At present, all the data layers are used in form of shapefiles (ESRI, 1998). The layers included in the map have an on/off option for display. The WebGIS interface also supports ancillary map graphics showing a legend, scale bar, display control buttons etc. The interface provides basic options for querying the PS points database for details of individual PS point records i.e., their location, topographic heights, ensemble coherence, yearly linear deformation and time series of linear deformation. The results of web-based visualization are shown for the case studies of Rotterdam and Delft City.

8.6 Case Studies: Cities of Rotterdam and Delft

The cities of Rotterdam and Delft are used as test cases for urban land subsidence analysis. Both these cities are comparatively densely populated, Rotterdam being one of the major European ports and Delft being a town of the famous technical university. For the Rotterdam PS-InSAR analysis, 79 ERS-1 and ERS-2 SAR images have been selected in the period 1992-2002. In case of Delft, 73 ERS-1 and ERS-2 SAR images over the city area are used. The time period of acquisitions range from year 1993 to year 2003.

The additional geo-information data sources included are GBKN building boundary map, Top10 boundaries of infrastructures and railroads, Top10 areas of land-use classes, and high-resolution aerial image of the city. The combination of these results is done using a MapServer Web-GIS. As explained in section 4, the Web- GIS visualization interface consists of various options of combining and querying the various data layers within a web browser. Some of the common combination examples could be joining, intersecting, selecting PS over a particular area etc.

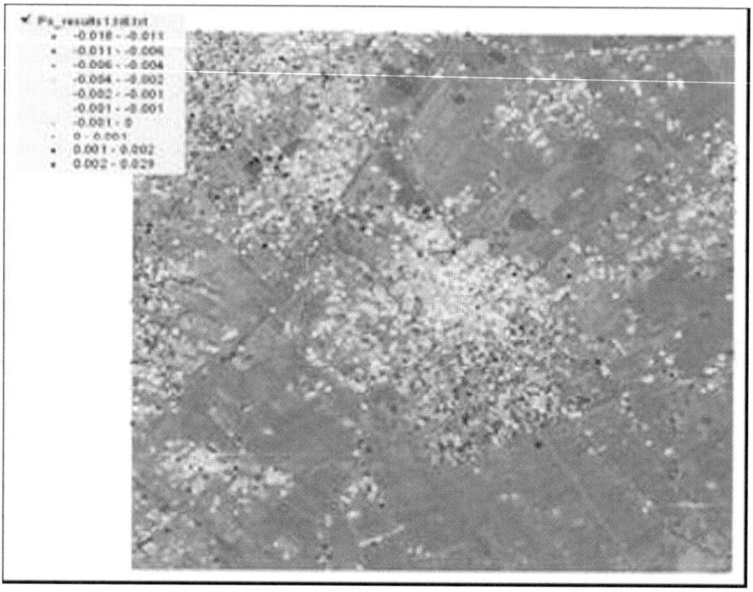

Fig. 1. PS points with radar image as background show displacement velocity from +20 mm/y (black) to –20 mm/y (white), relative to an arbitrary reference point

Figures 1 and 2 show the screen shots of desktop GIS and web based visualization interface respectively of PSI data combination for the Delft city. As seen from Fig. 1, the overall picture of all displacement profiles (PS) is useful to analyze the regional trend of land subsidence. The colour variation of points corresponds to the variation in linear displacement velocity of the persistent scatterers from +20 mm/year (blue) to–20 mm/year (red) with respect to the reference point. A clear trend of scatterers can be seen on roads, highways, and urban areas in Fig. 1. Observing the colour variation in Fig. 1, it can be deduced that the southern part of the city is comparatively more stable then the northern part. The probable reason for this instability in the northern part is partly due to the underground water extraction by a bio-chemical industry in the region.

To study the displacement profiles referring to local geographical features, the web-GIS interface is useful for a combined visualization of PS data. Figure 2 shows a screenshot of the visualization interface and the options of combining additional geo-information. When a required PS point is clicked with a mouse on the browser window, the attributive data and the deformation time series of the PS is displayed in a pop-up window. This functionality in web-based visualization is particularly useful when some party is interested in detailed behavior of the ground at particular spot(s).

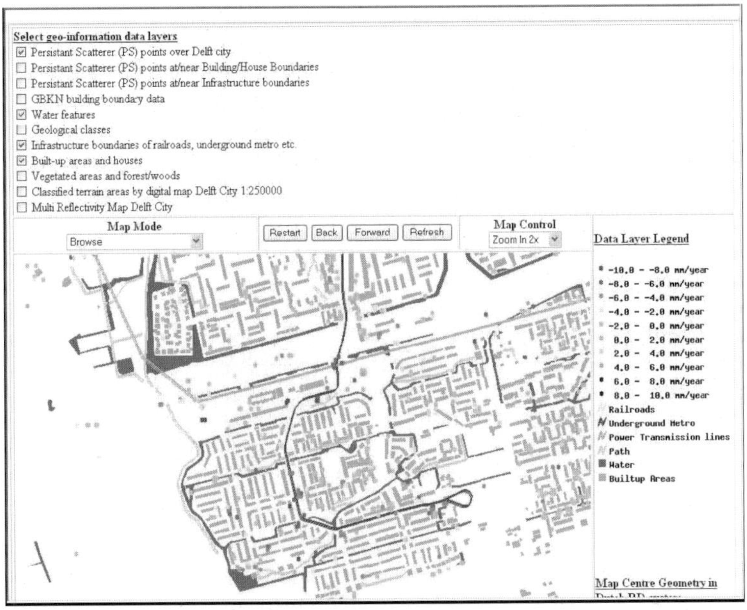

Fig. 2. Screenshot: Web-GIS functionality to show Delft PS data online with geo-data

Similarly, Figs. 3 and 4 show the PSI results over the city of Rotterdam in the Netherlands. Figure 3 shows the overall deformation scenario that consists of thousands of PS points. The yearly displacement values range from +10 mm (blue) to −10 mm (red). The supplementary data sources that are available for combined visualization are listed in the side bar of the browser window (Fig. 3). Interested individuals or organizations can interactively visualize the PS land subsidence data in their locality in reference to the land features they are aware of.

Figure 4 shows a zoom view of an area in the Rotterdam city that has a dense network of metro lines, bridges and tunnels using the web-GIS visualization interface. A trend of highly deforming points is observed over this area, and hence this information could be relevant for the city authorities.

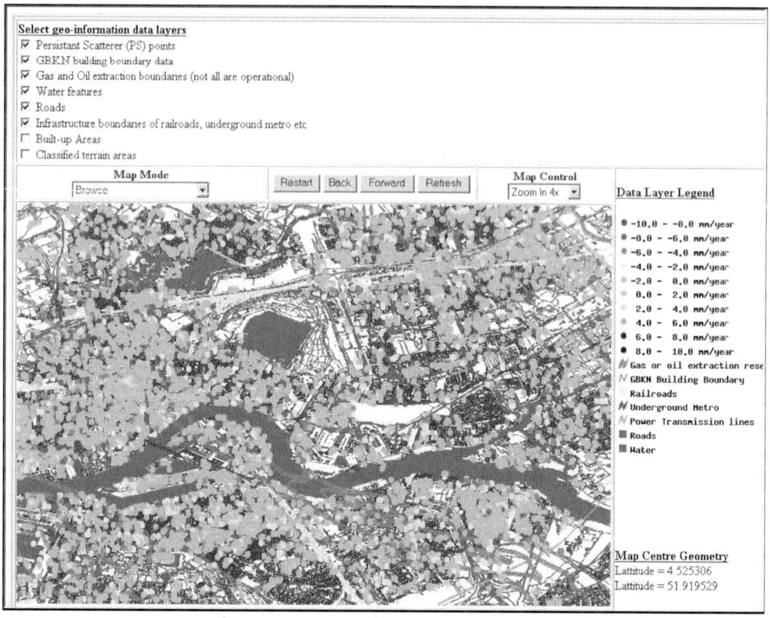

Fig. 3. Screenshot: Web-GIS functionality to show Rotterdam PS data online with geo-data

8.7 Conclusion and future research

Within this paper, we focus on the potential of land subsidence monitoring provided by PSI. Time series analysis of the radar data provides the opportunity to monitor slow movement of the earth's surface. The cost effectiveness of PSI is found to be very high in comparison to conventional geodetic techniques for subsidence monitoring. This research also investigates the options of including local geo-information to reason and validate the derived deformation rates as observed by the radar satellites.

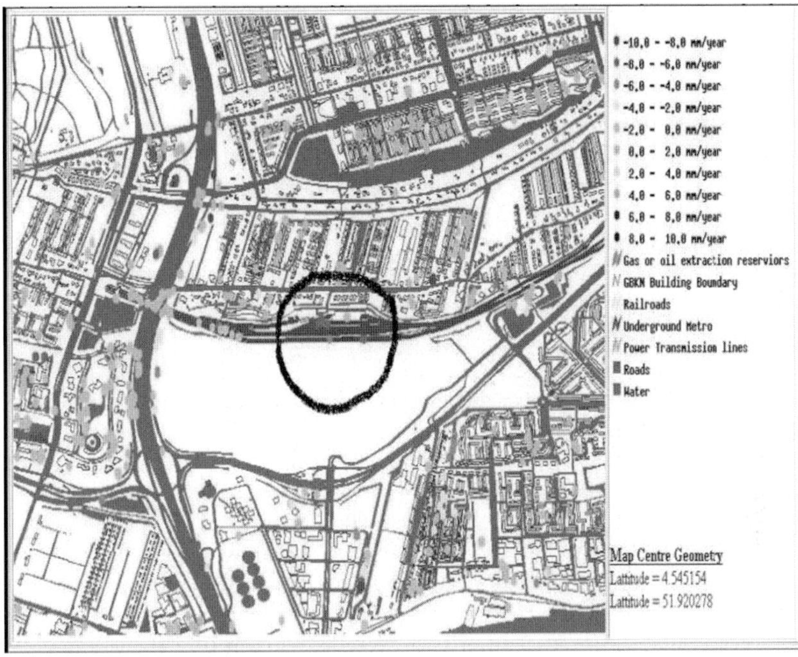

Fig. 4. Zoom view of Fig. 3 pointing high deformation points on a metro-line network

A variety of test cases are presented with the results of PSI analysis. A web-based dissemination and information-sharing platform is proposed to publish the results and make them understandable to non-expert users. The advantage of web-based GIS is that a user does not need to install expensive and dedicated GIS tools to explore geo-data. A simple web browser can be the means to visualize and interact with various data combinations.

Future research will focus on decomposition of the PSI deformation signal into various components such as, foundation instabilities, changes in ground water levels, thermal variations, subsidence related to soil compaction and geological phenomena. Solicitation of qualitative information by local experts via the web- GIS interface to add to the PS database is one of the tasks to be explored. With this research we aim to make Web-based GIS as a user-friendly tool to visualize, analyze, and interpret the complicated information about urban land subsidence.

References

ESRI Environmental Systems Research Institute, Inc. (1998) ESRI Shapefile Technical Description.

Ferretti A, Prati C, and Rocca F (2001) Permanent Scatteres in SAR Interferometry. IEEE Transactions on Geoscience and Remote Sensing, 39(1), pp 8-20

GBKN The Large Scale Standard Map of The Netherlands (2005) Official Website http://www.gbkn.nl/

Gehlot S, Verbree E, and Hanssen RF (2006) Dissemination of PS-InSAR Results for Improved Interpretation And Analysis. Online Proceedings: ESA FRINGE-05 workshop on Advances in SAR Interferometry from ENVISAT and ERS missions", Frascati, Italy.

Hanssen RF. (2001) Radar Interferometry: Data Interpretation and Error Analysis. Kluwer Publishers, Dodrecht

Massonnet, D and Feigl KL (1998) Radar interferometry and its application to changes in the earth's surface. Rev. Geophys. 36(4), pp 441–500.

OGC The Open Geospatial Consortium Inc. (2005) Official Website www.opengeospatial.org

Top10 Vector Map, Topographical Department of the Dutch Land Registary Office (2005) Official Website http://www.kadaster.nl

UMN Mapserver, University of Minnesota MapServer Documentation (2005) Official Website http://mapserver.gis.umn.edu/

9. Extending the Functionality of the Consumer-Grade GPS for More Efficient GIS and Mapping Applications

Robert M. Mikol

Abstract

This paper will present a methodology for extending the geospatial data collection functionality of the consumer-grade Global Positioning System (GPS) receiver for circumstances that previously required very expensive equipment, software and training. The approach presented here uses the waypoint name in a GPS receiver as an entire data record in the GPS database. This involves using a naming protocol based on a user-defined data dictionary. This procedure, when used with a relational database management system (RDBMS) and a geographic information system (GIS), can assist communities, organizations and agencies in the collection of high quality, accurate and sophisticated geospatial data under circumstances where funds and expertise are limited or nonexistent. Furthermore, the applications of this system are limitless. This paper shall discuss the applications of these protocols in several fields relating to disaster response and assessment; trail mapping; geospatial education and coastal resource assessment. Other applications shall be discussed or implied.

9.1 Introduction

The GPS is a ubiquitous tool on the planet today(Trimble, 2006). Its ability to pinpoint a person's position on the globe, and store information about that location has been one of the most significant contributions to society and human development in the latter Twentieth Century. In general, the GPS receiver is small, portable and inexpensive. As we settle into the Twenty-First Century we find that the standard GPS units are becoming smaller, less expensive, more accurate, more precise and more versatile with each new model released(El-Rabbany, 2006). They are found in nearly every manner of technical, recreational, military and civil geospatial data collection application.

Currently, the systematic collection of geospatial attribute data is limited to very expensive GPS receivers and specially trained personnel. The general procedure for collecting geospatial attributes with the consumer-grade GPS has required the use of 'add-ons', such as hand-held computers, laptops and palm pilots. These add-ons are frequently outside of the financial and technical reach of many communities, organizations and under funded agencies. The protocols and procedures presented here are designed to modify and merge existing methods and technologies in order to overcome these financial and technical barriers. The potential uses for the application of this protocol are unlimited. This paper attempts to describe a few of the applications where the protocol has been successfully deployed.

9.2 Methodology

The GPS provides four major data collection functions: (1) precise location through the collection of waypoints; (2) the precise time of data collection through the use of a date/time stamp during the waypoint collection; (3) a name for that location; and (4) tracks. These tracks are routinely collected by the GPS unit, which in turn can be displayed in a GIS and show an area surveyed or how and area was surveyed.

While items: One (1), Two (2) and Four (4) are self explanatory: item Three (3) requires the most discussion. The waypoint function of the GPS is the data collection engine for this naming convention protocol. It is in the waypoint name that the user is able to create a record of specific data attributes about that location. For example, it is common for the waypoint name to provide a location attribute such as 'HOME,' for a persons' domicile or an abbreviation like 'ST_J_LK,' for a place like St. John's Lake. In the GPS receiver's small database, the waypoint name is often limited to the number of character spaces available. Typically, the length of a waypoint name is limited to eight or ten characters. Another significant attribute of the waypoint name is that it also functions as the primary key field of the internal GPS database; as a result no two names can be alike. As we just mentioned, a common number of character spaces provided on today's GPS' are ten and the number of available alphanumeric characters are 36 (10 numerals and 26 letters). This alphanumeric feature allows the user to create a single character code for a word or description in an RDBMS table. Sometimes it is necessary to create a double character code of use two character spaces in order to record a numeric value greater than 10.

Using a data dictionary and a set of data structures for the waypoint name, even a casual GPS user can provide an organization with high quality, accurate and sophisticated geospatial data. A waypoint name, created with this protocol, is both a unique record for that location and the primary key value for the relational database (Getz, 1999). Once the waypoint name is parsed, a set of data fields are created with very specific attribute information about the location surveyed.

At a minimum, the waypoint name and subsequent parsed fields should describe: the GPS (including the operator); data id and a reference to that location. After this information has been entered, the possibilities are endless. In the event of an oil spill the additional data might include information on: shoreline type; presence of oil; vegetation; birds; animals; mammals and fish. An entire list of possible items need not be included in the lists to describe the shoreline, vegetation, etc. just those most common to the area. All of the fields created in the waypoint name and parsed out from the naming convention will be sorted out in the RDBMS (Kennedy 1996).

Prior to any field use, a data dictionary must be established. The nature of the survey will determine the format and content of the naming protocol and structure. For example a set of protocols for mapping out a trail system might only have five data entries in the name: GPS ID, Trail ID, Data ID, and Trail Attribute ID. The number of character spaces allotted for each data ID will depend on the number of items involved in the survey. If you do not expect to have more than 36 GPS receivers in your database, then one character space will be enough. The same is true for the number of trails to be surveyed. On the other hand, if the number of GPS units (or trails) can be expected to exceed the amount of 36 over the life of the project, then two character spaces should be allocated. Two character spaces will provide 1296 alphanumeric ID combinations.

So, a small trail mapping project might only have one character space allocated for the GPS ID, one space allocated for the Trail ID, two spaces assigned to the Data ID and one space allocated to the Trail Attribute ID. The waypoint name might look something like: '**270C4**'. This waypoint name, parsed out in the database provides the following information:

- **GPS ID:................2** (owner, make and model);
- **Trail ID:..............7** (St. John's Lake Trail, 23.8 km, moderate difficulty, average round trip speed 6 ~ 8 hours);
- **Data ID:................0C** (the 12th data entry for that trail);
- **Trail Attribute ID....4** (a bridge).

Connect this data into a GIS and you have a trail made from tracks whose symbology indicates moderate difficulty and a waypoint on that trail that indicates a bridge.

This same set of protocols and database design can now be modified for Search and Rescue (SAR): Trail ID might become Sector ID; and the Trail Attribute ID becomes an Item of Interest ID (an item of clothing, food wrapper or water bottle, etc. Other fields may be necessary such as an Obstacle ID indicating a river, ravine or other barrier that would force a new direction in a lost person's behavior. Another attribute might be a Facilitator ID, indicating an object that would encourage directional behavior, like a path, trail, road, cave, cabin or house.

Now imagine, 15 SAR personnel going out looking for a lost hiker. Each has a GPS receiver turned on, recording their tracks. Each member of the team also has a waypoint naming convention key that explains how to enter the waypoint name. As each person returns to base, the data in the GPS is downloaded directly into a database and immediately displayed in the GIS for review and analysis. In minutes and with minimal effort, a map document is produced showing the efforts, tracks and observations of 15 search and rescue team members.

9.3 Coastal oil spill model

Often when an environmental disaster strikes, the nearest local community is the first to know and the first to respond. Federal and state government resources, along with private contractors, though experts in their field, often take days to fully mobilize. During that time, precious information is lost through the absence of data. This is an opportunity where members of the community, with local knowledge of plants, wildlife, and terrain, can be mobilized to collect high quality, precision data with simple, consumer-grade GPS receivers. No one can collect this data faster than the local community with local knowledge. Additionally, by collecting this first response data, small communities now have a stronger position of involvement in the response and remediation of the emergency.

If a community is fortunate: then they have a flexible response plan in place and have had an opportunity practice that plan. The elements of a response plan are likely to proceed along the following outline:

- Purchase GPS units (or find out who has them)
- Develop a data dictionary for individual emergencies
- Develop a database (more precisely a Relational Database Management System or RDBMS)

- Store the Data Dictionary as a table in the RDBMS (this can then be printed as a report and distributed as the need arises)
- Simulate emergency situations and train personnel with the GPS receivers and GIS resources

Figure 1 is a representation of the oil spill data collection model. It is important to notice that several fields only allow for the specific choice of 33 values, where three input values are reserve characters for Not Present (0), Not Known (1) and Other (Z). This is particularly useful when a single character space is being used to identify an attribute. In many cases, 33 variables will be enough to identify the most common features that an agency or organization might be concerned about. Sometimes more than 33 variables are necessary. When this happens, two character fields can be combined to create a single data field, then the number of available codes increase with the number of possible alphanumeric combinations. In this case the number of possible combinations are 1296 (or 36^2).

Selandang Ayu Data Model (Emergency Response)

Charater Space Id Type	GPS ID	Data ID	Beach Type	Oil Present	Veg Type	Animal	Bird	Marine Mammal	Fish
Possible Choices	36	1296	33 + 3	2	33 +3	33 +3	33 +3	33 +3	33 +3

Some fields will require 3 reserved characters:
0 NOT PRESENT
1 NOT KNOWN
Z OTHER

... Leaving you with 33 characters for specific assignment

Fig. 1. GPS waypoint naming convention for the oil spill model

9.3.1 Data collection

Once an emergency has been identified it will be important to contact community members with knowledge of all aspects of the local geography, including the weather, vegetation and wildlife(NASAR, 2005). Next, pass out GPS receivers, the naming convention keys and a map, along with a track line or area to cover.

In the field, the waypoint data is collected according to established protocols. This means that each waypoint is given a name, and each character in the waypoint name corresponds to a specific attribute value in the data dictionary and that attribute value is specific to the location in the waypoint name. A written, hard copy record is made of the data point and this includes:

- Time of data entry
- Waypoint name
- Attributes identified

Additionally, tracks from the GPS are collected according to protocol. Most likely this will simply involve setting the track 'polling' rate or the running position data collection rate at a specific interval. This may be a standardized rate of 10, 20, 30 or 60 seconds, or it may vary with each GPS in order to distinguish one GPS unit from another.

9.3.2 Data integration

Upon returning from the field, each member of the data collection team needs to download the data from their GPS and load it into the relation database management system(Kennedy, 1996). This may be a direct download to the database, or it may involve more steps (Fig. 2). For the purpose of this protocol it is recommended that an organization use two more steps to ensure data quality. This is more likely to be necessary if you are using local members of a community who have had little or no training in emergency procedures or with GPS receivers and data collection techniques.

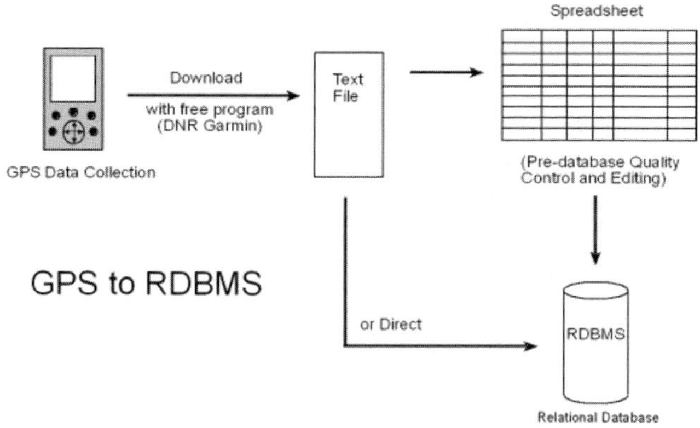

Fig. 2. This diagram illustrates the data pathway from the GPS to the RDBMS

Download the raw waypoint and track data from GPS to a text file. Next, label this file with the date of data collection (in reverse format), followed by the letter 'r' (indicating raw data), an underscore, and the first waypoint name (i.e. '060416r_0101011). The file now tells you the date of the data collection (16 April 2006) and by including the first waypoint in the file name you may be able to identify the GPS (possibly the user) and the area surveyed. The purpose for labeling the GPS files with the date first is so that these files will automatically sort chronologically in your computer's directory.

Next, import the file into a spreadsheet. This is a quality control step. It will allow for easy and efficient data checking; and if necessary data editing. Each user should be responsible for checking their own data for accuracy. If possible, this should be done with a member of the GIS team.

Next, import the 'clean' data (two files: waypoints and tracks) into the database. All of the waypoints can go into a single table for that project. The waypoint naming convention ensures that each waypoint can effectively be used as a primary key value. Additionally, the naming convention allows for extracting information related to a single GPS or survey area. The database administrator (DBA) may want to import track data into tables designed for each GPS. This would be the easiest and most effective way to distinguish one set of GPS tracks from another.

9.3.3 Data management

The RDBMS will need to have a set of reference tables and transaction tables. The example set listed below uses the Unalaska Trail Mapping Project (UTraMP) data model. The UTraMP model is particularly useful in that it is the easiest to understand and modify for single attribute acquisition.

To begin with, remember that the waypoint name is just a series of unparsed data values that will later be linked to several tables in a RDBMS. In the case of UTraMP there were four fields, using six character spaces in the data record: a GPS Id (consisting of two characters); Trail Id (one character); GPS Data Id (two characters, they were specific to that trail in a 'Parent-Child' inheritance relationship); and a Trail Attribute Id (one character). A waypoint name/data record liked like this: **02A0CG**.

- The first two values **'02'** indicated what GPS receiver was being used to collect the data.
- The letter **'A'** referenced the Trail being surveyed.

- The values **'0C'** indicated the sequence of the data point being recorded for this trail survey. This was the 12th data point to be recorded in this survey[1].
- The letter **'G'** indicated the trail attribute, in this case a 'G' referenced a Gate.

The relating tables in the RDBMS followed the general format listed below:

UTraMP Reference Tables:

GPS reference table (tbl_GPSId)
- cGPS_ID (GPS receiver Id number in the database)
- cGPS_Make (Manufacturer)
- cGPS_Model (Model of GPS)
- cGPS_SN (Serial number)
- cGPS_Owner (Owner or operator)
- cGPS_Name (Written on the case of the GPS, i.e. 'CRC-1,' in this case it meant the first GPS unit for that organization. Others were labeled CRC-2, CRC-3, etc.)

Trails reference table (tbl_TrailInfo)
- cTrail_ID (Trail ID number in the database)
- cTrail_Name (Name)
- nTrail_Length (the trail length in linear distance)
- nTrail_Time (the amount of time it takes to go out and come back)
- cTrail_Difficulty (on a scale of 1 ~ 3)

Trail Attribute table (tbl_WayPtsId)
- cTrail_AttributeID (This was the primary key value for the table and the waypoint key value for the naming protocol)
- cTrail_AtributeDescription (This was the description of the ID value, i.e.: 1 = Start of trail; C = Cabin. Note that the Attribute ID was a mnemonic to the description.

[1] The sequencing series is similar to the hexadecimal numbering system except that instead of recycling after the letter 'G', the value for 16. The letters move through the entire alphabet. The letter 'Z' has the numeric value of 35. The numeric value for the characters '10' is 36.

UTraMP Transaction Tables:

GPS Waypoints table (tbl_WyPts)

- cType (this field is information generated by the GPS unit and indicates whether the data is a waypoint or track point)
- cWypt_Name (a user created field, in this case, this is where the data record is generated and the attribute information for that specific location is stored)
- nLat (Latitude, generated by the GPS receiver)
- nLon (Longitude, generated by the GPS receiver)
- dtTime (Time and date, generated by the GPS receiver)

GPS Tracks table (tbl_Trax)

- cType (this field is information generated by the GPS unit to indicate whether the data is a waypoint or track point)
- nLat (Latitude, generated by the GPS receiver)
- nLon (Longitude, generated by the GPS receiver)
- nAlt (Altitude, based on user selected GPS datum, generated by the GPS receiver)

It is important to point out here that there is no data collected or generated that indicates which GPS receiver is generating the track data. This is where an organization may wish to put each set of a GPS receiver's track data into a separate table or add that identifying information while data is being checked in the spreadsheet program.

9.3.4 Data manipulation

Data manipulation is done through the use of queries: either by combining tables or queries of related data and/or running calculations on them. In this database, the primary form of data manipulation involves parsing the Waypoint Name field (cWyPt_Id) and 'joining' or linking the newly created parsed fields with the primary key fields of the appropriate attribute in the associated reference table.

Figure 3 illustrates the data flow for the Selandang Ayu Data Model for oil spills. The illustration is simplified from the actual RDBMS. In the actual RDBMS all data tables become queries before any relation is applied. This extra step is a common procedure done as a precaution to protect the integrity of the original data.

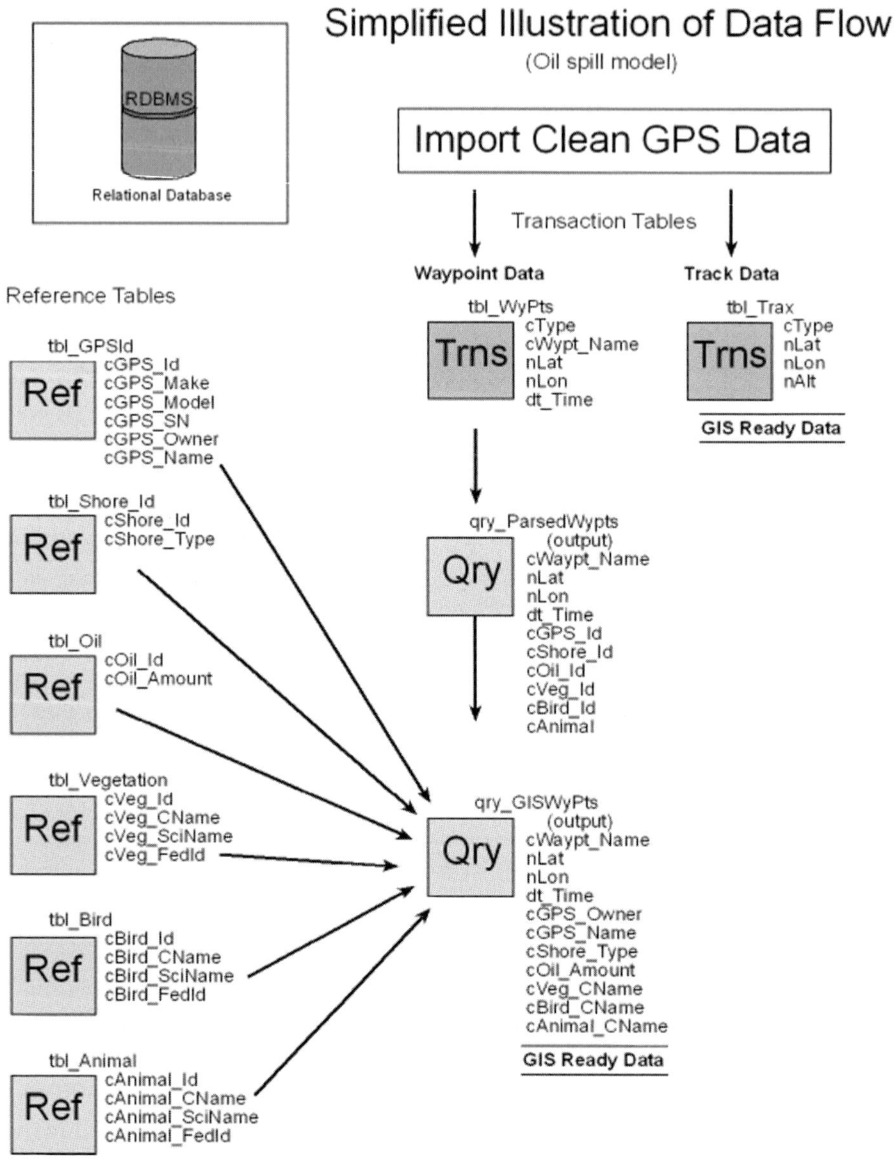

Fig. 3. This diagram illustrates a simplified database structure for the Oil Spill data model. Each of the reference tables are printed out as a key for use in the field as a guide for data entry.

9.3.5 Data analysis

At this point, the geospatial data is ready for visualization in a GIS. Now the management team can see what areas have been surveyed, who surveyed them and what was found. Most likely, you will have had to acquire or create our own base maps. You will have had to ensure that all of the maps and data are in the same projection with the same datum and you will have had to create symbology for the different attributes of your data(Iliffe, 2000).

At this point, emergency managers are ready to perform advanced analysis on the data. From here you can buffer points and lines; and use other tools that will allow you to clip, merge and intersect your GPS data with other map features (ESRI, 1998). Also, if this were a disaster, you could use advanced statistical calculations on the collected data in order to evaluate the nature and spread of a given contaminate. If this were a SAR operation, a view shed analysis could be performed from the elevation of the hiker and a maximum area of observation could be obtained.

A coastal resource assessment protocol has been investigated using this protocol. As of this time it is still in development. It was originally a modified version of the Oil Spill model, but going through a list of items and plugging them into a GPS was overly tedious and time consuming. Possibly, a better model, soon to be tested, involves limiting the number of items of concern to subsistence species or indicator species or both. In this case the data field would identify either a Boolean value of either presence or no presence, or presence and quantity.

9.4 Discussion

This waypoint/database model has been used and tested successfully in several environments. As previously mentioned, it was developed immediately following the grounding of the M/V Selendang Ayu, a large bulk carrier that broke up on the remote shores of Unalaska Island in the Bering Sea. Though the database was never used during the spill and subsequent cleanup, the idea grew and changed to meet the constant needs of collecting geospatial data where financial resources were limited.

Listed below are several data collection models and the lessons learned from each of them. In each case the final users of the data and the field data collectors were brought together in order to develop a database that could be reused far into the future with as little modifications as possible. Additionally, there was a recognition that if many users were brought into the development process, the final database design might become a standard and thereby have the capability of combining geospatial data from many different projects (i.e. stream data from different courses, different years, possibly even from different agencies). It was also recognized that one of the most important considerations of the database design included using federal codes (species, shore types, etc.), wherever possible. The purpose of this is two fold: one is to keep all aspects of the database as standard as possible; the other is to position the Waypoint Database Protocol for potential federal development funding, for ourselves and others who may work on this protocol.

9.4.1 The TRAIL model

The first production use of the database came with the community development of a trail map. Here, many data points needed to be collected in a short period of time. In this case several GPS receivers went out into the field each day. Each party carried with them a ‘key’ for the data dictionary describing the sequence of data input and what each character in a field stood for (i.e. 1 = Trailhead; B = Bridge: etc.). This simple model laid the foundation for other projects and data collection formats. The most important lessons to come out of this project were: keep the alphanumeric characters simple to remember for each of the data entries: B = Bridge; L = Lake; G = Gate, etc. If a letter in the alphabet did not have a corresponding real-world item, then it was left blank until one was appropriately found. Numbers and letters did not have to be filled up merely because they were the next character in sequence. If possible, keep the most commonly used characters as close to the default symbol of the GPS as possible. For example, when entering a waypoint name, if the GPS receiver’s waypoint naming function defaulted to a blank space that was centered between the number ‘9’ and the letter ‘A’ (like in the Garmin GPS receivers), then those alphanumeric characters may be the best choices for the most commonly used characters or for commonly used reserve characters.

9.4.2 The STREAM HABITAT model

A Stream Habitat model was designed for several graduate students at the University of Alaska, Fairbanks. In this model, 10 character spaces were used. The following is an outline of the design. Remember that date and time were recorded as a function of the GPS.

ID field **Number of Character Spaces**
GPS ID **1**
Stream ID **2**
Data ID **2**
Depth **2** (meters to tenths with no decimal, maximum 9.9m or entered as 99)
Temperature **1** (whole centigrade values, 0 degrees to 36 degrees: 0 – 9 as entered 10° entered as the letter A, 11° entered as B, etc.
Habitat ID **1** (33 possibilities with 3 reserved characters for not present, not known and other)
Obstruction ID **1** (33 possibilities with 3 reserved characters for not present, not known and other)

The data collection was arranged so that a field technician could follow a stream and collect baseline data at regular intervals. Additionally, if the need should arise to enter data between prescribed intervals due to a major change in habitat, depth, or obstruction, that new data would fit in seamlessly.

9.4.3 The EDUCATION model

Several workshops have been held using this method for teaching geospatial technologies, GPS and GIS. The target audience has been young adult in secondary school, grades 10 through 12. Here students go out and record a variety of items involving single character descriptions such as bird species, numeric values like the number of people on a corner; and Boolean values as in the presence or absence of garbage at these locations. In these workshop scenarios, students choose their own subject matter but organize it around the different data types: relational character values; direct numeric values; and Boolean values.

9.4.4 The COASTAL ASSESSMENT model

As mentioned earlier in this paper the Coastal Assessment Model is still under development. Several models have been tried, but the volume and complexity of the data available for collection make creating a standard data model difficult to design. However, the OIL SPILL model gets closest to a generic assessment model. The OIL SPILL Model was created with the idea that baseline data for that particular shoreline may not have been collected and archived. Beach type, slope, nesting sites and animal indicators are all apart of the collection process and corresponding reference tables.

9.4.5 The SEARCH AND RESCUE (SAR) model

Recently, we have started working with members of the Alaska Search and Rescue Association (ASARA) to develop a standard protocol that can be used state-wide. In addition to the immediate benefits that creating near real-time maps would have on the open investigation, it is believed that this geospatial database would provide a significant contribution to better understanding the behavior of persons subject to a search and rescue. The hope is to eventually run the database off of an internet server.

9.4.6 The Migration to Google Earth

Perhaps the most exciting development is the potential integration with Google Earth. If this develops as planned, then the rudimentary functions of collecting geospatial data and displaying the attributes of those data in a standardized format, on a standardized map will have taken a great step forward. This goal is not meant to substitute Emergency Management GIS but rather enhance it. It is truly emergency GIS when no other GIS is available.

Most organizations and small town public safety agencies do not have the money or expertise to own and operate a sophisticated GIS. However, if they have access to the internet, then it is possible to start bringing these GIS components together, geospatial data displayed on raster and vector maps.

The goal is to allow emergency managers the opportunity to upload GPS tracks and waypoints to an internet server which would then take the raw data, parse it out in the appropriate database and convert it into viewable distinct attributes in a KML file for display in Google Earth[2] (Google, 2007).

Google Earth has become the standard, web-based mapping service. It is a modest, but well functioning GIS in the public domain. In addition to having standard raster and vector maps available, users, are able to overlay images, measure distances and created vector files. The vector files are basic points, line and polygons. With a click of a mouse button, users can also email these data to other users, agencies and organization anywhere in the world for display on the same, standardized web-based maps. All the time, being able to keep the original GPS data available for a more powerful, dedicated GIS.

---

[2] KML (Keyhole Markup Language) is the markup language for the Google Earth program that displays geographic data.

The elimination of Selective Availability and the incorporation of WAAS and DGPS have greatly increased the accuracy and precision of the consumer-grade GPS. Expanding services on the Internet have also added a new potential to the functionality of the GPS receiver for use in emergency management and assessment. These new, tools have the potential to expand the resources of communities responding to challenging situations that require immediate geospatial information. Additionally, having this information and the knowledge that comes with it, as an emergency unfolds, can empower a community to make better decisions with the hope of a better outcome.

Acknowledgements

The author would like to thank Wendy Svarny-Hawthorne and the shareholders of the Ounalashka Corporation for their support in the UTraMP project, without them this protocol concept and method could not have been initiated. Further thanks go out to University of Alaska Fairbanks professors, Anupma Prakash and Michael Sfraga, both of whom gave their time and support to write up this project. Anupma was kind enough to read and comment on several drafts. Finally, Sisi Zlatanova from the Delft University of Technology and Tom Heinrich from the Geographic Information Network of Alaska, both encouraged writing up this project and helped to clean up several working drafts. Their kindness, patience and talents are greatly appreciated.

References

El-Rabbany, Ahmed (2006) Introduction to GPS, the Global Positioning System. Artech Publishing

ESRI, Environmental Systems Research Incorporated (1998) ESRI Technical Description

Getz, Ken and Litwin, Paul and Gilbert, Mike (1999) Access 2000 Developers Handbook Vol.1 Desktop Edition. Sybex Press

Google Inc. (2007) Official Website, GoogleEarth Tutorial, http://earth.google.com/userguide/v4/ug_kml.html

Iliffe, J.C. (2000) Datum and Map Projections for Remote Sensing, GIS and Surveying. Whittles Publishing, CRC Press

Kennedy, Michael (1996) The Global Positioning System and GIS: an Introduction. Ann Arbor Press

Lillesand, Thomas and Kiefer, Ralph and Chipman, Johnathan (2004) Remote Sensing and Image Interpretation 5th Ed. John Wiley & Son Publishers
NASAR, National Association for Search and Rescue (2005) The Handbook for Managing Land Search and Rescue. ERI Publications
Trimble Navigation Limited (2006) Official Website, GPS Tutorial, http://www.trimble.com/gps/whygps.shtml

Part 3: System Architectures for Access of Geo-Information

10 Interoperable Access Control for Geo Web Services in Disaster Management

Jan Herrmann

Abstract

In case of a disaster, the fast interoperable and secure exchange of spatial data is one of the key tasks. Access control plays an important role for disaster management in order to enforce restricted access to protected spatial data or to declare views on the relevant data for certain users/roles. Therefore, a powerful and interoperable language is needed to declare policies for operations on Web Services used, for example in disaster management. These polices have to contain rules that define which data can be accessed by a person/role through a certain operation (e.g. read, write) under given conditions (e.g. national or regional state of emergency). Different standards exist to establish access control for various requirements (e.g. OASIS's eXtensible Access Control Markup Language - XACML). However, no standard exists so far that allows the declaration and enforcement of spatial access restrictions needed for the geospatial problem domain. Based on the OASIS Standard XACML, a newly developed spatial extension to XACML called GeoXACML (OGC candidate specification) provides a possible recommendation on how to declare and enforce flexible access rights for spatial data in an interoperable way.

10.1 Introduction

The increased probability for the occurrence of disasters as well as the growing population density and the dependence of humans on existing systems or infrastructure implies that extreme events result growingly in disastrous situations. In order to keep the danger minimal to humans, their goods, and the environment, it is extremely important to optimize the single phases of disaster management (i.e. preparedness, mitigation, response, and recovery). To accomplish this goal, all parties involved must work together to create a suitable IT-infrastructure for disaster management (DM).

10.2 The profit of spatial data and geo web services in DM

Today, there is no doubt that spatial data, GIS and Geo Web Services, accessible over a network, are very useful in all phases of disaster management. The general requirement for a spatial data infrastructure (SDI) used in DM is to enable decision makers to get access to ideally all relevant data and services. Based on this data and services, the decision makers should be able to conduct their analyses and come to a sound decision.

Pilot projects such as Orchestra (Orchestra 2008) are proof that units involved in DM tasks can profit from many advantages by using SDIs based on standards. These projects also confirm that a SDI is an efficient way to handle the intense information exchange between the many different cooperation units at regional, national, and international levels. Unfortunately, these studies also show that, though the use of spatial data is established, the applied techniques to build SDIs for DM are still far away from an optimum.

The table below shows some crucial requirements for a SDI in DM (column 1). These needs can be seen as factors for success in order to allow the efficient use of spatial data and services in DM through SDIs. Also listed in the table (column 2) are the consequences derived from the requirements.

Table 1. Requirements for a SDI in DM and their consequences

Requirement	Consequence
Up-to-date data	Decentralized data management
Location independent, reliable and fast accessibility of data and services	Geo Web Services have to come into operation
Interoperability of data and services	Standards have to be used (e.g. from OGC or OASIS)
Availability of Metadata (e.g. SRS, date, QoS and ontology)	ISO 19115, OWL, etc.
Find/Bind mechanisms for data and services	Catalogue services
Integration and linking-up of existing systems and infrastructure	Use of standards and deployment of a security (and if necessary of a billing) framework
Secured access to data and services	Need for authentication, encryption, authorization, licensing etc.

In this chapter, the last requirement, the establishment of secured access to data and services within a SDI for DM is highlighted. Up to now, the need for security architectures for most systems or SDIs used in DM is identified, but this goal is often assigned to later phases of the corresponding projects (cp. Herrmann 2005).

10.3 Demands on a security framework in DM

First of all, one must understand why a SDI for DM needs to be secured. To illustrate this need, some use cases are introduced.

1. Recovery after a Hurricane. Imagine the coast of the US was hit by a Hurricane. The city council has established a Geo Web Service, e.g. a WMS providing maps showing the locations of mobile units supplying people with food and water. By this WMS, the citizens can be informed where the disaster recovery task forces are providing food and water at a certain moment. The city council uses the same Web Service to locate other mobile units (e.g. police cars). Now imagine some criminals who like to take advantage of the chaotic disaster situation. By monitoring the mentioned Web Service, they will always know the locations of all mobile police units. Being in possession of such information could cause organized despoilments. To solve this problem without setting up an additional service, the city council has to establish a simple access control system. Only if the user is authorized, he/she would be allowed to see the layer that shows the location of police cars.
2. Organizing an evacuation. A similar use case is the existence of a Web Service that indicates the location of the evacuation teams and the places where handicapped or older persons live. This information, that is essential for teams responsible for the evacuation process, could be dangerous if the wrong people get hold of this information (e.g. thieves). To avoid any risk, this sensitive data must be protected by an access control service.
3. Lack of information in case of a disaster. In many cases, data suppliers/vendors are in the possession of data which could be used in disaster management. Unfortunately, this data is often not available for the people involved in DM tasks. One reason for this inadequate situation is that data owners are worried about their data getting in possession of unauthorized persons and being misused. Because of the fear that they might harm their business by making their data available within an SDI for DM without security framework (especially without access control

system), they deny access completely.It is obvious that unnecessarily restrictive or even no access at all to existing and maybe relevant data is a drawback which can easily be solved by providing an access control system within a SDI. Thanks to an access control system, data providers could make their data available without any risk.

4. Legal and administrative directives. The UN demands from developed countries that disaster relevant data is transmitted early enough to affected countries (cp. van Oosterom 2005, p. 37). They appeal that the cooperation between developed and less-developed countries in the domain of disaster management is strengthened. This background implies that countries, on the one hand, have the duty to supply information and, on the other hand, they have to protect intellectual property and sensitive data.

To satisfy both requirements, one need access rules for the use of data in case of a disaster and for the use in disaster management respectively. This behavior of an SDI can again be achieved through an access control system built on top of secure communication channels. Clearly, a security architecture integrated in a SDI just offers the cornerstone, the technical solution to control the sharing of certain subsets of data within a determinate group of people. Complementary one has to establish binding administrative and legal regulations to optimize the exchange or accessibility of data needed in DM.

This listing of use cases could be extended at will. But even after the examples mentioned so far, one can already comprehend that security mechanisms (especially an access control framework) are crucial for SDIs used in disaster management.

To go into more detail, I would like to mention a poll regarding access control requirements for a service-based SDI accomplished at the Intergeo in 2002 (cp. Matheus 2005a). Based on this poll, the following requirements assuming an object-oriented data model have been addressed (cp. Matheus 2005b):

1. It is not sufficient to restrict the access to the entire service (coarse-grained restriction) as it allows an authorized user to access all provided data.
2. Context dependant access restrictions- It must be possible to declare access rights dependant on the context. Context in this case refers to the state of the system (time, number of accesses today, etc.)
3. Class-based restrictions- It shall be possible to restrict access to all resource objects (or parts of objects) of the same class. For a WMS, this requirement refers to the capability to restrict the map access for each individual layer or for a set of layers. For a WFS, this requirement refers

to the capability to restrict the access for a feature type, a set of feature types or certain elements, or attributes of a feature type.

4. Object-based restrictions- It must be possible to restrict access to individual objects (or parts of objects) of a class (e.g. access permit if owner_building = state).
5. Spatial restrictions- It shall be possible to restrict access based on the resource geometry if a certain spatial relation to a given geometry exists. In other words, features can be accessed if being, for example, within a given area or touching a given line.
6. It must be possible to declare positive and negative permissions in order to express restrictions such as, user A can read all features of type 'Building' but not the building named 'The White House'.
7. It must be possible to combine the context-based, class-based, object-based, and spatial permissions. As an example, user A can read all features of type 'Building' within the area of Washington D.C., but not the building named 'The White House'.

In the following sections, it will be addressed how one can fulfill all these requirements in an appropriate way.

10.4 Security

The flow of information between services and users/services in a SDI is exposed to many threats such as unauthorized access, wiretapping, modification, masquerade, Denial-of-Service, IP-spoofing, man-in-the-middle attack, etc. Furthermore, there is the increasing demand for security and protection of private or sensitive data and services usable over the internet. Therefore, desired characteristics of the communication are authentication, secrecy as well as message integrity, and authorization.

Apart from the authorization, the postulated characteristics are in the meantime achievable through techniques like SSL, IPSec, RSA, DES, MD5, Kerberos, X.509 certificate, SAML, etc. For this reason, the existence of a secured data channel including correct authentication can and will be assumed.

In contrast, the authorization problem was only rudimentarily solved so far, although it is an important pre-condition for the secured usage of Geo Web Services within an SDI. Sole action based access control on a coarse grained level like a file (e.g. Apache Web Server) or similar coarse grained concepts (access control in many database systems) are not sufficient for today's XML based applications (e.g. WFS-T). Due to the structured data model of XML, new authorization concepts are possible and needed for the content dependent regulation of accesses on a fine-grained level.

10.4.1 Access control

Based on a secure and authenticated communication channel, access control regulates under which conditions specific subjects (individuals or roles) may carry out certain actions (e.g. read, write, or calculate) on parts of the data. Different standards exist to establish access control for various requirements (e.g. OASIS's XACML or XRML). Like a detailed analysis (Herrmann 2005) has shown, the OASIS standard XACML is due to its powerful, advanced, and well-elaborated specification particularly suitable for the implementation of a flexible, scalable, distributed, and mature access control system for Geo Web Services.

Furthermore, this analysis demonstrated the suitability of rule based access control systems such as XACML to enforce the required access regulations.

In a rule based access control system, a set of rules defines when certain data sets can be accessed. A rule is defined by the following tuple: (effect, subject, action, resource, condition)

- effect = permit, deny, not applicable, indeterminate
- subject = the role, group or user to whom this rule applies
- action = for which action should the rule be evaluated (read, write, access)
- resource = which data should be protected by the rule (for XML data, for example, you might use an XPath expression to identify (discretionary fine-grained) the sub tree for which this rule regulates the access)
- condition = context and/or content dependant tests, that have to be fulfilled so that the rule gets evaluated

An example of a rule in Pseudo syntax:

(Permit, Eva, read, /Object/Building, if building is within Germany and alert phase = 2)

At this point it is necessary to mention that access rules can be defined referring to the request or to the reply to/of a Web Service. Depending on the use case (e.g. WMS or WFS, read or delete action), one can choose between one of the two options or use both

However, for SDIs and the geospatial problem domain respectively XACML is not powerful enough. In a SDI you primarily exchange, modify, or analyze spatial data. The particular characteristic of spatial data, the spatial reference and extent additionally permit a special kind of content dependant access control – spatial access control. Spatial access rules allow the definition of rule features (the area in which the rule is applicable) and tests for spatial relationships between the rule feature and the features for which access should be limited or granted. Through these spatial access rules, one can define if the access to a certain object is permitted or denied in dependence on the location and extent of the spatial data objects (e.g. features can be accessed if being within a given area or touching a given line).

Beside the fact that spatial access control offers a new way to declare access rights, it additionally offers an easy way to express complex and stable authorizations. Just imagine you like to allow access to buildings within 10 km of the boarder. Without spatial access control, you have to determine which buildings are within 10 km of the border and then you can code the IDs of these buildings into your access rules. In contrast, with a spatial access rule you can achieve this behavior by the one following statement:

(Permit, Eva, read, /Object/Building, if distance(acessed_Features, Border_line_string) <= 10 km)

Until a few months ago, no concepts existed on how spatial access control can be accomplished for Geo Web Services. Beginning from this point, we started our research project. The base for the project was the OASIS standard XACML (OASIS 2005) as well as SUN's XACML implementation (Proctor 2004). The result was a concept and a prototype implementation of an access control system in which you can define powerful, fine-grained, content-dependant, and spatial access rules.

The developed extension to XACML called GeoXACML (Matheus 2005b) enables the declaration and enforcement of access rights based on self defined GML features and tests for spatial relationships (cp. Matheus and Herrmann 2007). Thereby, the XACML extension points were used to augment the spatial functionalities.

Because GeoXACML is specific to the geospatial problem domain, standardization with OASIS is not possible. Therefore, a proposal for the spatial extension of XACML to GeoXACML was posted to the OGC community in order to provide a possible recommendation, how to declare and enforce access rights for object-oriented spatial data in an interoperable way (cp. Matheus and Herrmann 2007).

10.4.2 Access control based on XACML

An implementation of a SDI for DM is often based on services such as Web Services that communicate over a network. The deployed services support interoperable access and/or analytic functions by implementing open standards, among others, of the OGC. The most famous examples of these services are OGC's Web Map Service (WMS) and Web Feature Service (WFS).

The figure below shows a possible service-based SDI with access control based on XACML. The components that belong to the access control system are shaded in dark grey.

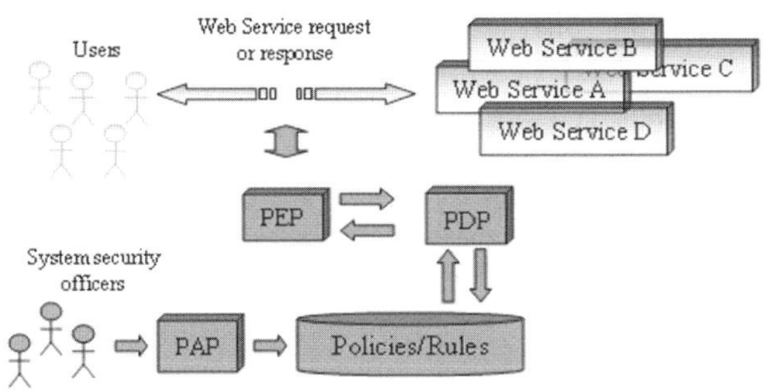

Fig. 1. A service-based SDI extended by an interoperable access control system

Every time a subject carries out an action on a Web Service, the access control system has to be activated in order to find matching rules, to evaluate these rules and to return the access decision.

In detail, the PEP (Policy Enforcement Point) is responsible to start the access control process by submitting requests for access decisions to the PDP after each request of a subject for an action on a Web Service.

In turn, the PDP (Policy Decision Point) identifies the relevant rules and evaluates them. If there is more than one matching rule, the effects of the matching rules have to be combined by a certain algorithm (e.g. deny-overrides, first-applicable, etc.). In this context, the structuring of policies and rules within policies plays a decisive role. The rules and policies are managed through the PAP (Policy Administration Point), which can be a Web Service itself.

The result of the PDP, the answer to the requests for access decisions, is then submitted to the PEP. The PEP afterwards handles the original Web Service request or answer accordant to the result of the evaluation of the whole policy. At this point it is possible to configure the PEP in two ways.

In case of a violation with the policy, the PEP either gives back the intersection of accessible and requested data or a simple deny.

GeoXACML

The proposed spatial extension of XACML, named GeoXACML, is based on the open standard XACML, version 2.0 of OASIS (OASIS 2005).

This section shows the most important elements that were added to XACML by the predefined extension points. These elements form the new possibilities and functions of GeoXACML. For a detailed description of GeoXACML or XACML, the reader is referred to Matheus and Herrmann (2007) and OASIS (2005).

10.4.2.1 Geometry data types for rule attributes

The added data types allow the definition of arbitrary simple features in one of the encodings as specified in the GeoXACML candidate specification (cp. Matheus and Herrmann 2007).

Table 2. Geometry data types for rule attributes

Geometry types	
Point	MultiPoint
LineString	MultiLineString
Polygon	MultiPolygon

10.4.2.2 Spatial rule functions

The functions of GeoXACML allow the validation of two geometries. For the exact semantics of the functions, the reader is referred to the GeoXACML candidate specification (cp. Matheus and Herrmann (2007) and OGC (1999) respectively).

Table 3. Spatial functions

Spatial functions			
Disjoint	intersects	boundary	intersection
Touches	equals	convexhull	union
Crosses	contains	centroid	area
Within	Is_within_distance	difference	distance
overlaps	buffer	symdifference	length

10.4.3 Example

The following example shows how to express spatial access rules in GeoXACML for a WFS (syntax simplified). Please note that the examples do not cope with the full functionality of GeoXACML. In particular, no combination of context-, class-, object-based and spatial permissions is used in this example.

The spatial access rule defined below denies access to data of building features if the area of the building intersects with area A, which is defined by the polygon (20,380 40,200 … 20,380).

```
<Rule Effect='Deny'>
 <Target>
  <AnySubject/>
  <Resource>
  <!-- this rule is evaluated if Building nodes are in the WFS response -->
   <ResourceMatch MatchId='regexp-string-match'>
    <AttributeValue DataType='string'>
/ResourceContent/FeatureCollection\[\d+\]/featureMember\[\d+\]/building\[d+]
    </AttributeValue>
    <ResourceAttrDesignator DataType='string' AttributeId = resource-id'/>
   </ResourceMatch>
  </Resource>
  <AnyAction/>
 </Target>
 <Condition FunctionId='intersects'>
  <Apply FunctionId='spatialInformationSelector'>
<ResourceAttributeDesignator DataType='string' AttributeId = 'resource-id'/>
<AttributeValue DataType='string'>/Geometrie/gml:Polygon
  </Apply>
  <!-- the second parameter of the intersects function, the area A -->
  <AttributeValue DataType='spatial'>
   <gml:Polygon srsName= '.../gml/srs/epsg.xml#4326'>
    <gml:outerBoundaryIs>
     <gml:LinearRing>
      <gml:coordinates decimal='.' cs=',' ts=' '>
20,380 40,200 240,140 360,260 340,380 200,400 100,400 20,380
      </gml:coordinates>
     </gml:LinearRing>
    </gml:outerBoundaryIs>
   </gml:Polygon>
  </AttributeValue>
 </Condition>
</Rule>
```

The target definition above implies that the rule only gets evaluated if any subject conducts any action on a 'Building' feature. In case where there are 'Building' XML elements in the answer document of the WFS, the spatial access restriction in the condition element will be evaluated (i.e. Does the building_geometry intersects with area A?). If the accessed buildings intersect with the rule area A, the accordant building elements in the WFS response will be deleted by the XSLT processor. Afterwards, the PEP can decide whether the user gets the modified WFS response or a simple access denied message.

10.5 Conclusion

As discussed, access control makes a contribution to the improvement of SDIs used in DM. The needed requirements for access control in DM can mainly be achieved with the OASIS standard XACML. XACML enables the implementation of an interoperable, modular, scalable, and distributed access control system for Geo Web Services supporting the declaration of fine-grained, content dependant authorizations for read and modify actions. By using the extension GeoXACML, spatial access rules are additionally possible.

One can conclude that GeoXACML meets the requirements for an access control system usable in DM. The developed concept and prototype was tested and validated in test beds with Web Feature and Web Map Services and has proved to be ready for operation. At the end, I would like to mention two further application domains of GeoXACML.

A GeoXACML-based access control system can be used to control the selective distribution of data. In many articles, one can read about the need of a well-organized information flow in DM so that the right persons get the right information in the right format in time to make the right decisions. In this context, the access control rules are used to define the views of the data.

Another interesting area of application of GeoXACML is the possibility to express and enforce price and billing models in GeoXACML. Even though this use case of GeoXACML is more business related, it is nevertheless worth mentioning, as in many cases the SDI established for DM might or must be used for the sales of data and services at the same time.

References

Herrmann J (2005) Spatial Access Control for Geo Web Services. Diploma thesis, Technical University Munich. http://www.geobranchen.de/images/produkte/GEOdownloads/herrmann_entwicklung.pdf

Matheus A (2005a) Declaration and Enforcement of Access Restrictions for Distributed Geospatial Information Objects. Ph.D. thesis, Technical University Munich. http://tumb1.biblio.tu-muenchen.de/publ/diss/in/2005/matheus.pdf

Matheus A (2005b) GeoXACML – a spatial extension to XACML. OGC Discussion Paper. https://portal.opengeospatial.org/files/index.php?artifact_id=10471

Matheus A, Herrmann J (2007) OGC Candidate standard specification: Geospatial eXtensible Access Control Markup Language. http://portal.opengeospatial.org/files/?arti-fact_id=25218&version=1

OASIS (ed) (2005) eXtensible Access Control Markup Language (XACML) Version 2.0. Committee Specification. hhttp://docs.oasis-open.org/xacml/2.0/access_control-xacml-2.0-core-spec-os.pdf

OGC (ed) (1999) OpenGIS Simple Features Specification for SQL Revision 1.1. http://www.opengis.org/techno/specs/99-049.pdf

van Oosterom P et al (eds) (2005) Geo-information for Disaster Management. Berlin, Heidelberg

Orchestra (ed) (2008) The Orchestra Project. http://www.eu-orchestra.org/

Proctor S (2004) Sun's XACML Implementation Programmer's Guide for Version 1.2. http://sunxacml.sourceforge.net/guide.html

11 Spatial Data Infrastructure for Emergency Response in Netherlands

Henk Scholten, Steven Fruijter, Arta Dilo and Erik van Borkulo

Abstract

During the crisis management, several organizations coordinate their emergency work based not only on well-defined policies and procedures (product of careful preparation) but also on the outcomes of the decision-making process. Decision-making is a highly complicated process in crisis situations. Good support in decision-making when disaster occurs is of critical importance to react accurately, fast and effectively. Good decision-making helps to control damage, save lives and resources, and reduce unwanted consequences of a crisis. Spatial Data Infrastructure (SDI) is increasingly considered a critical aspect of decision-making in disaster management. This paper presents our concept for an appropriate SDI (open-standard generic platform) assisting in administration, analysis of data to predict development of a crisis event and accordingly advise for changes of plans

11.1 Introduction

The first hours after a disaster happens are very chaotic and difficult but perhaps the most important for successfully fighting the consequences, saving human lives and reducing damages in private and public properties. In these first hours, the good estimation of the current situation is of particular importance. A large number of questions (e.g. where is the emergency situation, how to get there, how many casualties, what kind of information do we have from there), are pending for answering.

Additionally to the unclear situation, many different actors from different sectors (fire brigade, police, paramedics, municipalities, etc.) with, sometimes, contradictory requirements are involved in managing the situation. Since many of the tasks they have to perform are rather different than their daily work routines, a good cooperation and collaboration between different teams is of critical importance.

There are numerous underlying obstacles that complicate an adequate crisis response (Borkulo et al 2005, Diehl and van der Heide 2005, Kevany 2005, Neuvel et al 2006, Ospina, 2006, Winter et al 2005):

- Lack of good communication between sectors or actors in crisis response. Different sectors operate with particular systems; therefore the exchange of information between all actors in the crisis management becomes difficult.
- Diversity of systems in one sector. It is quite common practice to use various software in different regions or groups of regions. Realising this problem, for example, the Dutch police has already initiated a large project SHERPA aiming at standardisation of the geo-information provision of all the 25 police regions in the Netherlands.
- A variety of narrow specialised systems for only emergency situations.
- Lack of appropriate platform (user interfaces) for data exchange. The interfaces usually serve tasks and activities of a specific sector (police, fire-fighters, etc.)
- Insufficient standardisation of processes and protocols, especially for data exchange.
- Difficulties in exchange and integration of various data. Resolution, accuracy, updatedness are only few of the challenges.
- Lack of information about the 'information'. Often is unclear where certain information can be found and how reliable it is.
- Management of information from the field (reports, images, video, dynamic data, etc.). Usually this information is stored as unstructured files, which is problematic for systemised analysis, or it is not stored at all.
- Access to existing data is in general very slow.

New systems have to be developed that allow different rescue units to operate together in any critical situation. One of the main challenges is in the spatial domain, i.e. definition of well-defined standardised services for discovery and exchange of information. Such services are closely related also to the development of Spatial Data Infrastructure (SDI) at all levels (local, regional, national and international) for support of disaster management. Initiatives for SDIs are in progress at many levels all over the world, for example INSPIRE in Europe (www.ec-gis.org/inspire). Those have to be further enriched to be able to serve emergency sector. Large international projects, for example, ORCHESTRA (www.eu-orchestra.org), OASIS (www.oasis-fp6.org) and WIN (www.win-eu.org) are already reporting results of their research in this area.

Three aspects of an SDI are especially critical for the success of an emergency response system: generic services that would be available for all the actors (on the field and in the commando centre) in a disaster situation; integration of information coming from different sectors, and management of dynamic information; appropriate interfaces for different end-users.

The services developed within this research are context-oriented with respect to the 19 types of disaster as specified in the Netherlands (Diehl and Van der Heide, 2005). User interfaces allow for both request of data and submitting (in-situ) data to the system and are built on systems currently in use in the response sector.

Many kinds of information are needed in crisis response. These are existing data produced by different organisations, and dynamic data collected during a crisis from the different sectors involved in crisis response. Integration of data coming from different sources is an important aspect. Maintenance of dynamic data received from the field is still rather underestimated. Most of the information exchange is currently performed via the telephone.

This paper presents concepts and ideas for the development of an SDI for disaster management. Next section presents the state-of-the-art in the development of SDI. Section 3 discusses the services that are being developed. Section 4 elaborates on graphic user interface used in the moment and demanded by the users. Section 5 concentrates on information management, and a data model for the management of in-situ information. Last section concludes on expected added value of the developments. This research is part of the project 'Geographical Data Infrastructure for Disaster Management' (www.gdi4dm.nl) funded by the Dutch Research and Development Programme 'Space for geo-information'.

11.2 SDI in emergency response

SDI has been considered an important aspect of sharing data for years (Abrue et al 2000, Cattenstart and Scholten, 1999, Scholten et al 1999) Currently there are many international (Fabri and Weets 2005) and national initiatives for building SDI (INSPIRE, NEN, etc.). The reports for the EU member states show that a lot of progress is being made towards establishing European SDIs. There has also been progress toward a process of data harmonization as well as to common systems and data models in disaster response and prevention (ORCHESTRA, OASIS, GMES). However, the large number of activities and projects funded by EU shows that the SDI is far from reality.

From a technical point of view most of the technology that is required for access and exchange of spatial information is available as standards for implementation (e.g. OGC Web Services, SFS, GML), as concepts (e.g. OGC Abstract specifications for open distributed management of geographical imagery) or in process of development. It is worth mentioning that third phase of the OGC Web Services has been completed at the end of last year. Within these phase participants worked on several relevant topics for disaster response such as common architecture, sensor web enablement, Geo-decision support services (GeoDSS), Geo-Digital Right Management (GeoDRM) and Open Location services (OpenLS). Many extensions of existing standards are proposed for further discussions. For example the OGC OpenLS is extended toward indoor tracking (www.opengeospatial.org).

For the communication between the systems in this architecture, the use of web services and XML can be considered the state of the art. On the side of the clients, there is a great diversity ranging from rich clients based on web browsers over classic workstation applications to mainframes with terminals. These clients cover all the use cases for spatial data that so far exist and are based on highly heterogeneous programming and data models.

There are also many contributions from scientific projects. It should be mentioned many approaches for reaching semantic interoperability, search engines based on ontologies, security for agents and web services, as well as processing, indexing and serialization algorithms in the field of computer science. The topic of semantic interoperability is regarded as central and projects with strong industry integration like WIDE8 (IST 2001-34417) as well as those with a relation to metadata interoperability (like the INVISIP9 project, IST-2000-29640) are in progress.

11.3 Generic services

Our major objective within this project is to offer services that solve many of the obstacles that are now faced by emergency response systems. These services aim at improving communication between different actors involved in crises response. The services are part of an infrastructure that integrates and facilitates access to various information, existing data and data coming from the filed. It offers metadata – information about 'the information', exchange of data, etc. The focus is largely on developing context-aware services.

Defining and implementing an SDI is traditionally done by creating application profiles, where the data model behind the data infrastructure is leading. Generic services based upon open standards can then be developed to make exchange of data possible. The major disadvantage of this approach is that the development of services can be complex and time consuming.

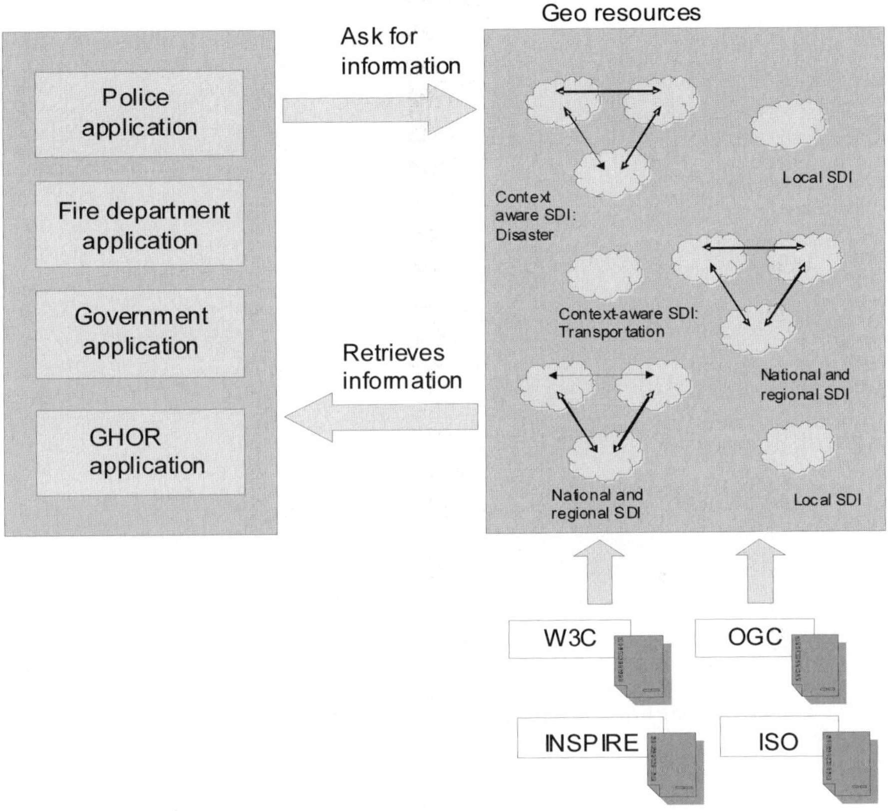

Fig. 1. An overview of context-aware SDI for disaster management.

A different approach is defining and implementing an SDI within the context it is primarily used (Fig. 1). For this project the context is defined with respect to the 19 different disaster types. In this approach, the use scenarios within the current disaster type prevail for the implementation of services within the SDI. Such services only provide access to the data infrastructure that is relevant for the current disaster type. This approach assures to maintain the focus within a certain context, as well as that application designers/architects to be able to use the defined and implemented services within their known specialization.

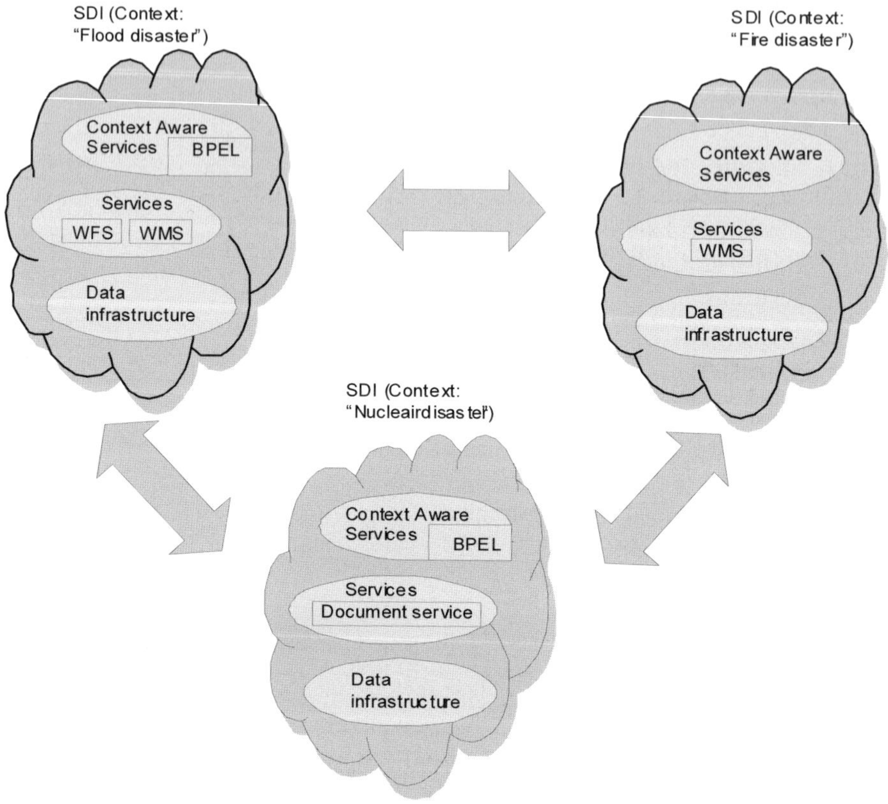

Fig. 2. An example of SDI for the types of disasters: Flood disaster SDI, Fire disaster SDI and Nuclear disaster SDI.

Within data driven (or syntax-aware) SDI's the applications are directly communicating with the services, which make them very dependant and aware of the syntax being used by services or data infrastructures. In context driven SDI's high-level functions within the active context are exposed to the application making the dependency of the underlying services less relevant (Fig. 2). These functions often consist of a series of calls to various services within the context-aware SDI. This approach is often referred as service chaining. This chain of calls to the various services ismodelled after the workflow designed to be used during emergency situations

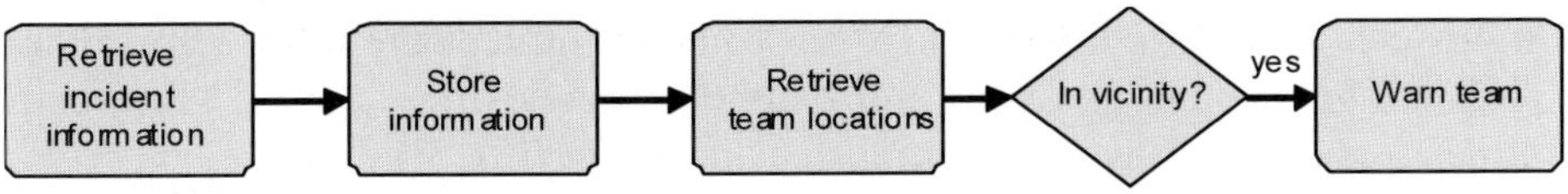

Fig. 3. Workflow of an emergency process.

Figure 3 shows an example of a workflow during an emergency situation. Information of an incident is retrieved (e.g. by phone or electronically). This information is stored for example by writing it on a paper or whiteboard. In order to warn the appropriate teams, the teams which are in the neighbourhood of the incident are selected and warned if necessary (e.g. by using radiotelephone).

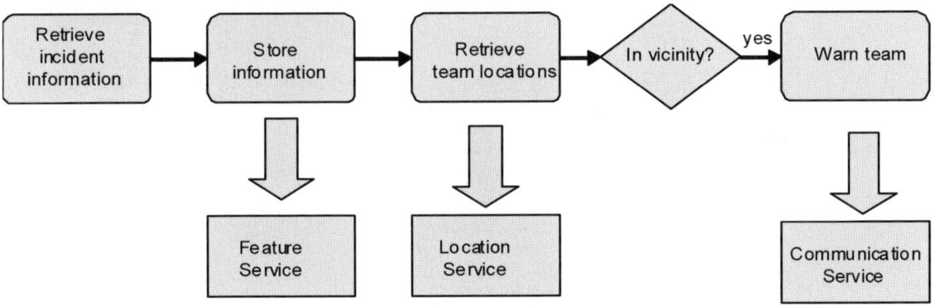

Fig. 4. Workflow of an emergency process implemented using service chaining.

Figure 4 shows the same example where the actual actions are implemented by a chain of calls to services belonging to the SDI. The Feature Service stores the information in the database described in chapter 5. The Location Service (e.g. an OpenLS implementation) is responsible for the tracking and tracing of the teams. Finally the Communication Service takes care of sending the information to the teams. Techniques such as BPEL (Business Process Execution Language) are available to model and implement the service chaining. BPEL is an XML based language describing among others the succession of calls to the different services (also called orchestration).

11.4 Graphic user interface

The design and development of an appropriate Graphic User Interface (GUI) is yet another challenging task in disaster management. GUI is usually very specific with respect to a particular user, type of device used and functionality to be offered. In disaster management these specifics are even more extreme, due to different user backgrounds, stress, time pressure, fatigue, etc. Therefore the interface has to be intuitive, easy for use and functional. This project analyses and extends/modifies the GUI of Multiteam (Figs. 5 and 6) (www.multiteam.info) and VNet (Fig. 7).

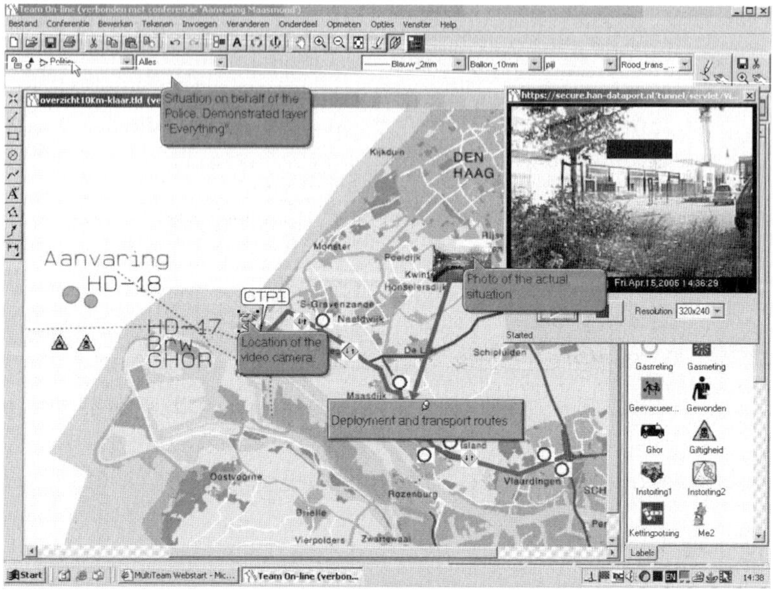

Fig. 5. Graphic user interface of the system Multiteam.

In both systems the different responding agencies in the crisis response (Fire Service, Para medic, Police, municipalities, other special units) can access the system and give the location of their mobile-units (using special symbols) or mark important areas e.g. those not accessible to the Public. The user of the system can send e-mails and request different maps as a background. The systems differ slightly in their functionality and access to the information. While Multiteam has a quite large local database with information, the concept of VNET (Diehl and Van der Heide 2005) is accessing distributed information (stored within the organisations responsible for their own service delivery). In both systems, however, (spatial) analysis is not available yet. The only existing functionalities are map overlay and visual inspection. Simulations (as discussed in flood risk management) are not available at the moment. In addition, compatible communication systems are being developed to improve communication during imminent floods.

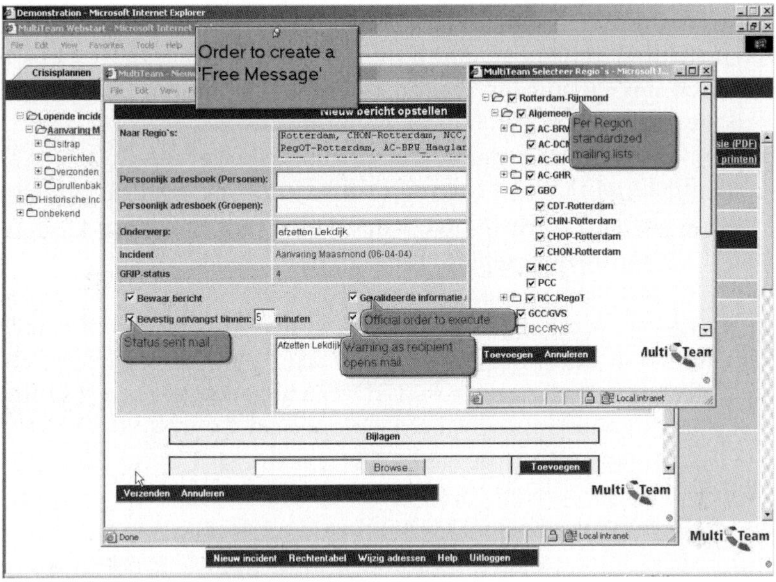

Fig. 6. Interface for e-mail communication.

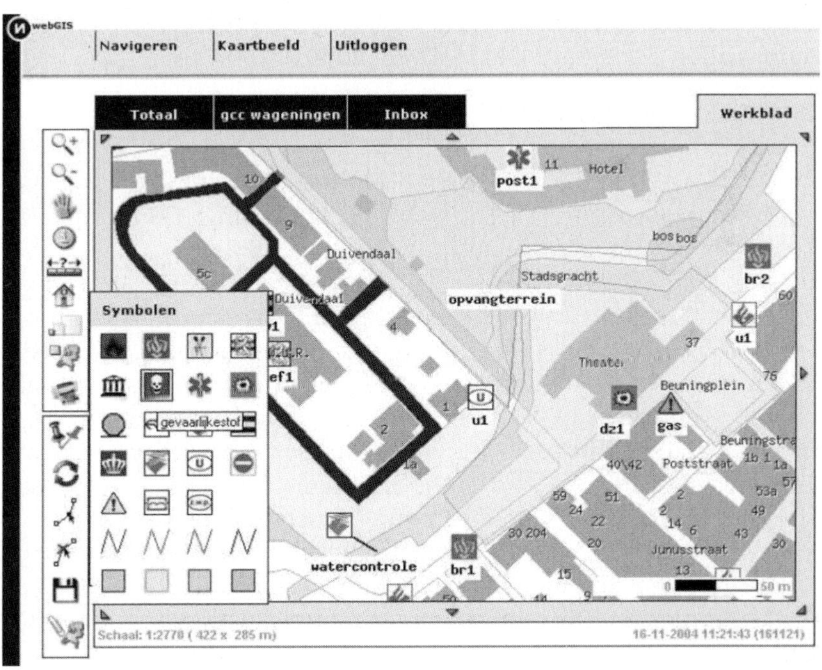

Fig. 7. Graphic user interface of VNet.

The GUI interface in emergency response should be intuitive, and simple, yet providing the most appropriate information with respect to the user, the used device and the emergency of the situation. Furthermore the access to the data can be dependent on the location of the object, the process, the task and the particular organisation (Capelleveen 2005). It should be also remembered that the behaviour of people changes in critical situations. In this respect, different GUI interfaces is to developed for mobile users and decision-makers in commando centres.

3D display and new visualisation environments are considered as well. Various groups are working already in providing 3D visualisation (Berlo et al 2005, Branco et al 2005, Bodum et al 2005, Kolbe at al 2005, Kwam and Lee 2004). It is well known that 3D visualisation gives a lot of advantages but puts big challenges to the developers of the system. For example, the 3D visualisation has to be very close to the real view. In contrast to maps, where a lot of symbols are used, 3D view should convey by realism and not by abstraction. Special techniques have to be used to focus attention to the most important information. For example, usage of a textured building amongst shaded ones. 3D models may be represented with plenty of details but in most of the cases this may lead to distraction. In this respect it is very important to keep the balance between important and fortuitous information (Zlatanova and Holweg, 2004).

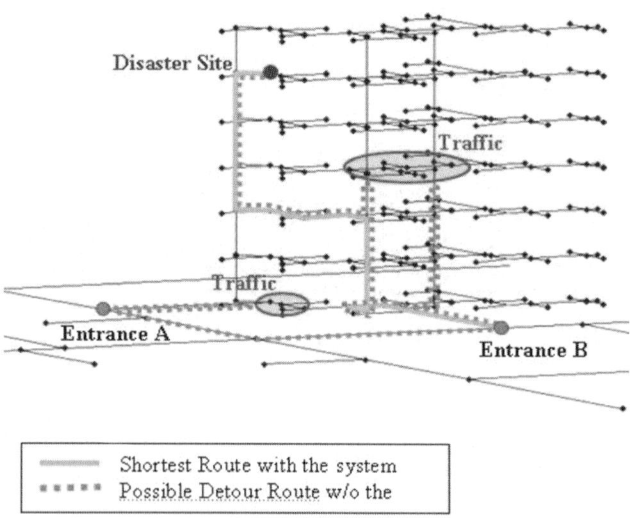

Fig. 8. Navigation routes in a building (from Kwam and Lee, 2004).

3D representations and displays can be critical for indoor routing and navigation (Kwam and Lee 2004, Meijers at al 2005). Considering the complexity of modern buildings and the great number of people that can be inside buildings, it is often challenging to organise a quick evacuation. Very often serious problems such as huddle, trample, inaccessibility of exits, etc are observed. Despite all the currently available means (alarm signals, evacuation plans, illuminated signs, etc.), three serious deficiencies are observed (Pu and Zlatanova, 2005): 1) lack of appropriate indoor geo-information about the structure and the available exits of the building, 2) lack of dynamic information about the current situation and 3) lack of flexible means for evacuation instructions Therefore have investigated possibilities to employ interfaces that dynamically evaluate the possible escape routes. The evacuation routes can be computed using network models (see next section) and the visualization as 3D graphics or appropriate images (Fig. 8) can be transferred to the client application.Very often 3D models are not available, but still a sort of 3D effect can be obtained. (Verbree et al, 2004) have studied the applicability of panoramic images, a CycloMedia (www.cyclomedia.nl) full-colour images or Cyclorama's. The intention of CycloMedia is to cover the Netherlands with cycloramas to support all kinds of geo-information systems and services. All objects and locations are registered systematically from the public roads within cities, districts, etc. The cycloramas are recorded by a very special fisheye lens with a vertical view of 30 degrees below the horizon.

Fig. 9. Cyclorama's, augmented with utility information (from Verbree et al, 2004).

Being geo-referenced, the images can be used for integration with various data. Actually, the cyclorama can function as the 'background layer' to display information labels, guiding directions, and position hotspots of other cycloramas in a given neighbourhood. The cycloramas can be integrated in any GIS system, as a clickable object either at the location where the image is taken from or at given addresses. By this, the user gets a far better impression of the surrounding objects and situation. Figure 9 illustrates a cyclorama image augmented with underground utility information. The cable data are taken from a 2D digital map and transformed in the projection of the image. Such an approach is readily applicable even for the restricted resources (screen, memory, bandwidth, etc.) of mobile devices. The image processing can be performed on a server and the augmented image can be streamed to any type of wireless devices.

This project extensively explores new emerging tangible technologies. Such technologies provide realistic visualization, human-computer interaction, help in situational awareness, and allow for cognitive mapping and collaborative decision making (Scotta et al 2006). Figure 10 illustrates the touch table to be used in the research.

Fig. 10. Geodan tangible user interface table (from Scotta et al, 2006).

The Diamond Touch Table is a front projected table. This means that the image is displayed with a beamer hanging above the table. The touch signal is recognized with an array of antennas embedded in the touch surface (www.merl.com). This technique allows not only to determine where the table is touched but also who is touching the table.

11.5 Management of information

The information needed during emergency response consists of existing information, e.g. topographic data, network utilities, etc., and information collected at the moment of disaster, the *in-situ* information. The integration of existing information is done through the use of ontologies (Xu et al.

2007). In this section we concentrate only on the management of in-situ information, which is an important aspect of the project. This information has to be stored and processed in order to predict the development of a crisis event. The stored information of past crisis events can be further analyzed and used for the prevention phase.

A DBMS system is used for the management of in-situ data, for several important functionality such a system offers: reliable management of large data sets, multi-user control on shared data and crash recovery, automatic locks of single objects while using database transactions for updates, etc. Nowadays several commercial and non-commercial DBMSs offer support for spatial data types: Ingres, Oracle, Informix, IBM DB2, MySQL, PostgreSQL. Some DBMSs (e.g. Oracle Spatial) even support several different spatial models (geometry, topology, network, LRS). The network model is quite appropriate for commutation of evacuation routes combined with 2D/3D visualisation since the geometry of the objects can be stored together with the nodes and links of the graph (Pu and Zlatanova 2005, Meijers et al 2005). These DBMS systems support only 2D spatial data types with their 3D coordinates. Full consideration of the third dimension may require new data types (Zlatanova and Stoter 2006). If needed, 3D topologies (Oosterom et al 2002, Zlatanova et al 2004) at database level and corresponding operations will be investigated. 3D topology may be required to extend the spatial functionality with operations such as 3D routing, and consistent update of data. Some DBMS systems support the temporal aspect, but the spatial and temporal indexing is treated independently. Good spatiotemporal support needs an integrated approach, consisting of proper data types, operations and specific indexing (and clustering) structures (see also Laurini et al, 2005).

A critical question is the selection of the DBMS, as well as the choice between commercial or open source. Important aspects to be considered are the support for different data types, appropriate for handling different kind of information coming during an emergency; support for the temporal component; extendibility with new types and functions. It is expected that information collected during an emergency is of very different nature. Besides various sensor information such as images (terrestrial, aerial), videos, laser-scan (also terrestrial and airborne), large amounts of textual data, audio, etc have to be managed. Most of the information is dynamic; therefore the temporal component is critical. The choice between open source or commercial DBMS is driven from pragmatic reasons. On the one hand, an open source, freeware DBMS (e.g. PostGIS) may have benefits in large area devastating disasters (similar to East Asia Tsunami or the hurricane Katrina) when existing infrastructure is destroyed and a commando centre has to be set up in few hours. On the other hand, many organizations in the Netherlands have already commercial DBMS, such as Oracle Spatial.

Oracle Spatial and PostGIS offer the most from the functionality we need. Many units from the emergency sector, e.g. municipalities, have already their data in Oracle. For this reason we have chosen to use Oracle Spatial. We have built a conceptual model for the in-situ data, and have translated it to Oracle Spatial tables, including constraints, indices, validation checks, and new data types. The rest of this section describes a part of the model (Fig. 11), which covers the information used during emergency with dangerous substances (three disaster types).

Figure 11 shows the tables that store the information, and relationships between the tables. Table `Incident` stores information about the location of a crisis event, the time it started, its extent, type of the disaster, etc. Two other tables, `Sectormal` and `Gasmal`, store information about the possible extent of the incident, calculated from meteorological data. A series of processes is initiated to handle a crisis event. Each process is well defined, and has a specific objective, which determines on which kind of disasters the process will be needed. For example, in case of emergency with dangerous substances, measurements should be performed in the affected area, and the process called 'Observations and Measurements' (Borkulo et al 2005) has to start. The table `Process` stores information about each process according to the Dutch emergency response procedure: the process type, the incident for which it is initiated, the time the process started and the time it concludes. Table `DM_User` stores information about personnel of disaster management sectors, and table `UserInProcess` stores information about which person is involved in which process and the time period of this involvement. Table `Measurement` stores the result of the 'Observation and Measurements' process. Several measurements may be performed during such a process. The measurement team performs measurements for the checked items, and reports the results in a corresponding form. Table `Measurement` contains the information from both forms: the specialist form defining the tasks, and the measurement team form giving the results of measurements. The `Team` table contains information about teams performing measurements. `TeamMember` table keeps track of persons forming a team, and the period of time they are in the team. The model shown in Fig. 11 does not contain all validity checks and indices; it is though complete in terms of primary key constraints (`PK` symbol) and foreign key constraints (`FK`).

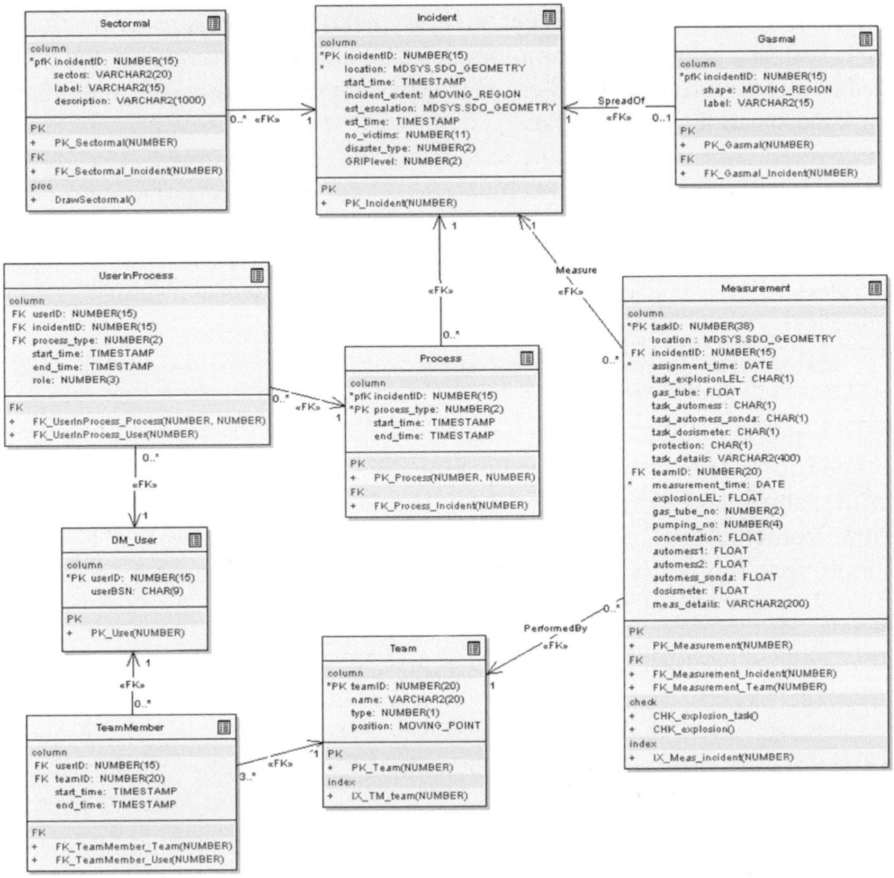

Fig. 11. UML model of Oracle tables holding in-situ information in case of emergency with dangerous substances.

The Oracle Spatial data type `MDSYS.SDO_GEOMETRY` is used for spatial data, e.g. location of an incident that is a point, or the incident estimated extent that is an area. To store spatiotemporal data we use nested tables of Oracle. An area that changes in time, e.g. the extent of an incident, is stored as a list of pairs (area-shape, time). We have created a new data type `MOVING_REGION`:

```
/* Types for a moving region instance, and complete moving region */
CREATE OR REPLACE TYPE MRegionInst AS OBJECT (
    meas_time TIMESTAMP,
    region_geo MDSYS.SDO_GEOMETRY
);
/
CREATE OR REPLACE TYPE MOVING_REGION AS TABLE OF MRegionInst;
/
```

Another spatiotemporal type, `MOVING_POINT`, is created in a similar way. It stores information for a location that changes in time, e.g. the position of a team in the field. These new types are used in the SQL statements for table creation, for example:

```
CREATE TABLE Incident (
    incidentID NUMBER(15) CONSTRAINT PK_Incident PRIMARY KEY,
    location MDSYS.SDO_GEOMETRY NOT NULL,
    start_time DATE,
    incident_extent MOVING_REGION,
    disast_type NUMBER(2) CONSTRAINT REF_DT REFERENCES Disast_Type,
    GRIPlevel NUMBER(1),
    est_time TIMESTAMP,
    est_escalation MDSYS.SDO_GEOMETRY,
    no_victims NUMBER(11)
)
NESTED TABLE incident_extent STORE AS IncExtent;
```

We use Oracle nested tables functionalities to store spatiotemporal data in the tables, and to read these data. The SQL statement below creates a view from the `Incident` table that transforms spatiotemporal types to simply spatial types (via un-nesting), which can be easily used by services.

```
CREATE VIEW IncidentExtent AS
    SELECT i.incidentID, i.location, i.start_time, e.*
    FROM _Incident i, TABLE(i.incident_extent) e;
```

11.6 Conclusions

The standardised services developed within this project facilitate the work of emergency sector in a number of directions. We expect that the systems will allow for a better preparation to disaster management, because of the standardised operations able to be exercised in daily work. On the other hand, the temporal data collected from previous disasters can be analysed and used for prevention.

It will be possible a better handling of the first period of an emergency, achieved by providing targeted information rapidly and by collecting relevant information from the field that helps coordinating operations efficiently. This period, called the golden hour, often determines the success of the entire operation, for instance as concerns the ability to save lives.

The integration of command-and-control with field operations can be significantly improved especially the coordination between forces, teams and organisations that participate in emergency management. The nature of the project is such that this impact can be achieved as concerns communication and data management.

The handling of emergency will also be improved, because information from multiple sources and databases will be possible to access in order to provide the emergency sector with all insights that can be realistically obtained in support of life saving and protection of material assets. This can prevent that vital data will not be usable simply because the data handling technical issues.

The visualisation of requested information will be improved with respect to the specific needs of the users and therefore provided in formats (e.g. images, vector, movies) that are more compatible with the environmental conditions during the emergency.

This 'definitely' will result in a better support to rescuers, based on information provision and coordination, which can on the one hand lead to faster and more effective operations and in the other hand ease the burden of stress on the operations theatre.

However, even 'definitely' needs to be evaluated and proven. As has been stated in the report of the National Research Council (2007) there are a number of factors that will influence the success of a map. Part of the project will be the development of evaluation procedures which will be used during the planned large disaster management exercises. Based on these outcomes we will be able to judge the role of the SDI

References

Abrue, J., Scholten, H. J., Eijnden, van den B., Gehrels, B. 'Combining Spatial Metadata Search-Engines with Web mapping' in: GIS, Journal for Spatial Information and Decision Making, Vol. 13, October 2000, pp. 8-11.

Berlo, M. van, van Rijk, R. and Buiël, E. A PC-based Virtual Environment for training team decision-making in high-risk situations, in: Proceedings of the 2nd International ISCRAM Conference April 2005, Brussels, Belgium, pp. 195-201.

Bodum, L., Kjems, E., Kolar, J., Ilsøe, P.M. and Overby, J. GRIFINOR: Integrated Object-oriented Solution for navigating real-time 3D virtual environments. in: PJM van Oosterom, S Zlatanova & EM Fendel (Eds.), Geo-information for disaster management, Springer Verlag, Heidelberg, pp. 937-949.

Borkulo, E van, Scholten, H. J., Zlatanova, S. and van den Brink, A. Decision making in response and relief phases, in: PJM van Oosterom, S Zlatanova & EM Fendel (Eds.), Geo-information for disaster management - late papers, pp. 47-54.

Branco, P., Escalada, C. and Santos, R., 2005. Real-time 3D environment simulation applied to the disaster management field: our experience, in: PJM van Oosterom, S Zlatanova & EM Fendel (Eds.), Geo-information for disaster management, Springer Verlag, Heidelberg, pp. 425-441.

Cattenstart F. J. and Scholten, H. J. 1999, Towards OpenGIS Systems, The Vector Data Storage Component Evaluated, In: Interoperating Geographic Information Systems, A. Bekovski, K.Brassel & H. Schek (eds), Springer Verlag, Berlin, pp 163-176.

Capelleveen, E. van, 2005, Information management boost command & control, in: P.J.M. van Oosterom, S. Zlatanova & E.M. Fendel (Eds.), Geo-information for disaster management, Springer Verlag, Heidelberg, pp. 979-995.

Diehl, S. and van der Heide, J. 2005, Geo Information Breaks through Sector Think, in: P.J.M. van Oosterom, S. Zlatanova & E.M. Fendel (Eds.), Geo-information for disaster management, Springer Verlag, Heidelberg, pp. 85-108.

Fabri, K. and Weets, G. 2005, ICT for Environmental Risk Management in the EU Research Context, in: P.J.M. van Oosterom, S. Zlatanova & E.M. Fendel (Eds.), Geo-information for disaster management, Springer Verlag, Heidelberg pp. 51-56.

Kevany, M. 2005, Geo-information for disaster management: lessons from 9/11, in: P.J.M. van Oosterom, S. Zlatanova & E.M. Fendel (Eds.), Geo-information for disaster management, Springer Verlag, Heidelberg, pp. 443-464.

Kolbe, T., Gröger, G. and Plümer, L. 2005, CityGML: Interoperable access to 3D city models, in: P.J.M. van Oosterom, S. Zlatanova & E.M. Fendel (Eds.), Geo-information for disaster management, Springer Verlag, Heidelberg, pp. 883-899.

Kwan, M-P. and J. Lee, 2005. Emergency response after 9/11: the potential of real-time 3D GIS for quick emergency response in micro-spatial environments. Computers, Environment and Urban Systems, 29: 93-113.

Laurini, R., Servigne, S., and Noel, G. 2005, Soft Real-Time GIS for disaster monitoring, in: P.J.M. van Oosterom, S. Zlatanova & E.M. Fendel (Eds.), Geo-information for disaster management, Springer Verlag, Heidelberg, pp. 445-479.

Meijers, M., Zlatanova, S. and Preifer, N. 2005, 3D geoinformation indoors: structuring for evacuation, in: Proceedings of Next generation 3D city models, 21-22 June, Bonn, Germany, 6 p.

National Research Council, 2007. Successful Response Starts with a Map: Improving Geospatial Support for Disaster Management. Committee on Planning for Catastrophe: A Blueprint for Improving Geospatial Data, Tools, and Infrastructure,198 pp. National Academies Press

Neuvel, J., van Borkulo, E., Scholten, H. J., Zlatanova, S., Beugelink, G. and van den Brink, A. 2006, De geografische dimensies van risicomanagement, Definitiestudierapport, Bsik project, 71 p.

Oosterom, P. van, Stoter, J., Quak, W. and Zlatanova, S. 2002, The balance between geometry and topology, , in: Advances in Spatial Data Handling, 10th International Symposium on Spatial Data Handling, D. Richardson and P. van Oosterom (Eds.), Springer-Verlag, Berlin, pp. 209-224.

Ospina, M., 2006, What did GIS programs learn from Katrina? In Proceedings of 25th Urban Data Management Symposium, 15-17 May, Aalborg, Denmark, 6.29-6.38

Pu, S. and Zlatanova, S. 2005, Evacuation route calculation of inner buildings, in: P.J.M. van Oosterom, S. Zlatanova & E.M. Fendel (Eds.), Geo-information for disaster management, Springer Verlag, Heidelberg, pp. 1143-1161.

Romao, T. I., Dias, A. E., Camara, A. S., Buurman, J. and Scholten, H. J. 2000, Decision Support Tools for Coastal Zone Management: Providing Decision Makers with Useful Information Through Multidimensional Visualisation Techniques', in: International Journal of Development Planning Literature, Vol. 15, No. 1, January-March, 2000, pp. 33-40.

Scholten, H.J., LoCashio, A.J. and Overduin, T. 1999, Towards a spatial information infrastructure for flood management in The Netherlands, In: Special features in coastal conservation 3, eds J. Visser & R. Misdorp, Opulus Press, Uppsala, pp 151-160.

Scotta, A., Pleizier, I. and Scholten, H. 2006, Tangible user interfaces in order to improve collaborative interactions and decision-making, in Proceedings of 25th Urban Data Management Symposium, 15-17 May, Aalborg, Denmark, pp. 9.27-9.39

Verbree, E., Zlatanova, S. and Smit, K. 2004, Interactive navigation services through value-added navigation CycloMedia panoramic images, in: Proceedings of Sixth International Conference on Electronic Commerce, 25-27 October, Delft, The Netherlands, CDROM, 10 p

Winter, S., Iglesias, G., Zlatanova, S. and Kuhn, W. GI for the public: the terrorist attack in London, Geoinformatics Magazine, No. 7, Oct/Nov 2005, Vol. 8., pp. 18-21.

Xu, W. and Zlatanova, S., Ontologies for Disaster Management Response, In: J. Li, S. Zlatanova and A. Fabbri (Eds.); Geomatics Solutions for Disaster Management, Heidelberg, pp. 185-200

Zlatanova, S., 2005, Crisis designs, Geospatial Today, May-June 2005, Vol 4, Issues 1, pp. 30-36.

Zlatanova, S. and Holweg, D. 2004, 3D Geo-information in emergency response: a framework, in: Proceedings of the Fourth International Symposium on Mobile Mapping Technology (MMT'2004), March 29-31, Kunming, China 6 p

Zlatanova, S., Fabbri, A., Li, J. Geo-information for Disaster Management: Large scale 3D data needed by Urban Areas, GIM International, March 2005, Volume 19, Issue 3, pp. 10-13.

Zlatanova, S. and J. Stoter, 2006, The role of DBMS in the new generation GIS architecture, Chapter 8 in S. Rana & J. Sharma (Eds.) Frontiers of Geographic Information Technology, Springer, pp. 155-180.

12. Geocollaboration in Hazard, Risk and Response: Practical Experience with Real-Time Geocollaboration at Québec Civil Security

Charles Siegel, Donald Fortin and Yves Gauthier

Abstract

The pertinence of outputs from geographic analysis and airborne- and space-based sensors to hazard recognition, risk analysis and emergency response is well established. They are contributors – remotely-sensed or analytical information is not sufficient by itself. Useful application of geospatial data requires information from multiple sources, first during analysis and then to provide geographical context. The resulting product is useful to emergency management (EM) specialists, planners and other decision makers. Even then, however, these users may not have the capability to apply the information effectively. The problem is one of delivery: how to deliver the information rapidly and accompany it with the expertise required for use.

The delivery problem is particularly acute in decision making in the EM response phase. During response, the utility of geospatial information is related to the time it takes to acquire, process and analyze data, and then make it pertinent in the decision environment. Furthermore, these steps are not simply linear – analysis of the geospatial information may have to be adjusted, and very rapidly, using *in situ* measures and observations.

One way to resolve the problem is to add a georeferenced, real-time collaboration environment to the software and procedures used to produce geospatial information. This combination allows adjustments in analysis and interpretation, along with the input of expertise from multiple domains. In this paper, we describe applications of geospatial information with software for live Internet geocollaboration in civil security operating in the Québec Ministry of Public Security. The illustrations come from real emergency management situations.

12.1 A familiar map with new information at the SCQ

Under the Québec provincial government, central and regional umbrella bodies coordinate emergency management activities. These organizations, the OSCQ and ORSCs (*Organisation de la sécurité civile du Québec* and *Organisations régionales de la Sécurité civile*), are made up of the provincial departments and para-governmental bodies that have specific missions under law or the provincial civil security plan. A department of the Ministry of Public Security (MSP), the DGSCSI (*Direction générale de la Sécurité civile et de la Sécurité incendie*), has the mission of coordinating overall provincial action and information flow in case of civil emergencies. Practically speaking, the DGSCSI furnishes the backbone of civil emergency management in Québec (as "Québec Civil Security", *Sécurité civile du Québec* [SCQ]). The department runs the provincial and regional emergency operations centers (EOCs), and organizes and coordinates the umbrella organizations. The system of civil emergency management in Québec is organized a bit differently than in many other North American states, provinces and federal governments, where EM response is based on Incident Command principles.

Despite organizational differences, the objectives and missions of emergency management organizations (EMOs) are similar: recognition of hazards and reduction of risk, planning and preparation for emergencies, response and recovery. EMOs achieve these missions through collaboration with many other actors – other government ministries, departments and agencies, para-governmental organizations and specialists from the private sector or the academic world. Collaboration gives the EMO a pool of expertise far larger than the organization itself can provide (Siegel et al., 2005; Committee on Intersections between Geospatial Information and Information Technology 2003).

In general, the information shared among all these actors should be georeferenced (or at least "georeference-able") so that we can combine it and put it into geographic context for end-user applications or operations. One of the ways of thinking about these interactions is to imagine three spheres of collaboration: technical, operational and strategic (that is, planning or decision-making), which themselves interact based on a common view or map (Fig. 1). Remote sensing and geoanalytical information products are one of many possible sources. They must be used and interpreted in the same geographic context as other inputs to operational coordination and decision-making. Ideally, they should also be integrated into the typical practices of the EMO.

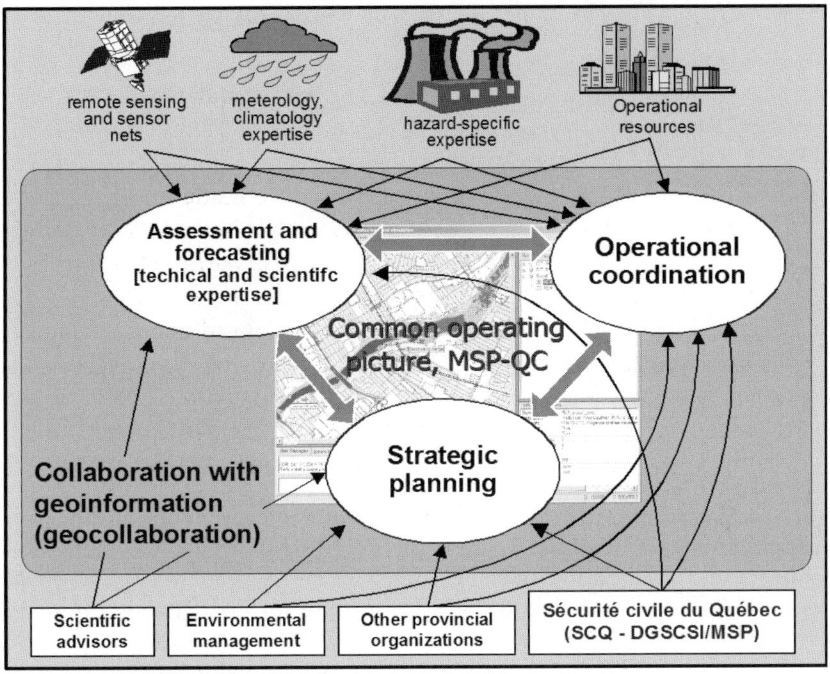

Fig. 1. Geocollaboration in emergency management at Québec Civil Security (*Sécurité civile du Québec* [SCQ]).

What is this context? Since we are dealing with people in society, we need to see where the population is, where the infrastructures are, what the physical conditions are, and so on – operational or situational information has to be considered in that framework. Domain specialists, including those working with remote sensing or geoanalytical information, need to explain the new information they contribute. In emergency management in Québec, that means providing a map view combining base information from federal and provincial geodata stores with the information provided by subject matter experts. The emergency management team then has a familiar map view allowing it to recognize, quickly and easily, the implications of new information (Fig. 2).

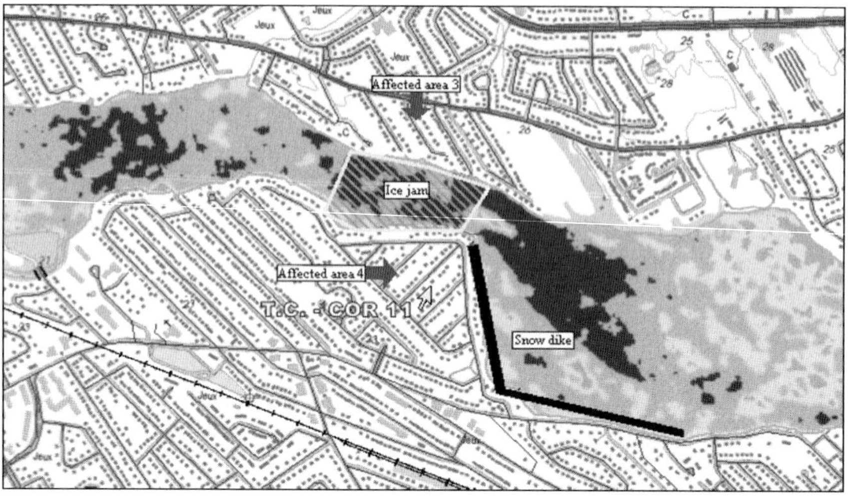

Fig. 2. Specialized information put into context for emergency management: river ice mapping based on Radarsat imagery (produced by INRS-ETE Remote Sensing Laboratory) and information from observers in the geographic context of WMS layers from Québec (MRNF) and Canadian (Centre for Topographic Information, Natural Resources Canada) sources. This illustration is taken from a screen capture in GeoConference®, the geocollaboration tool used by the SCQ.

Having assembled the map information in a usable form, we then need a means of providing it to the EM collaborators so they can use it together. The team is likely to be spatially distributed, and gets together using a both face-to-face meetings and telephone conferences. The map must be part of those team meetings. The SCQ has implemented an off-the-shelf Internet collaboration tool called *GeoConference®* to provide a real-time collaboration environment in those meetings.

An architecture for map-based conferencing. The architecture of GeoConference is straightforward and optimized to support the conferencing aspect of collaboration. The system consists of three tiers, which can reside on different, Internet-connected computers (Fig. 3):

- Conferencing clients,
- The GeoConference server,
- DataProviders (geoinformation connectors), which furnish map layers to the server for the conference session map background.

Participants access sessions using either stand-alone thin clients based on the Microsoft® .NET framework or Web-browser-based clients based on asynchronous JavaScript and XML (AJAX) Stand-alone clients can connect to the GeoConference server using any of the following protocols:

- Simple object access protocol (SOAP) over HTTP,
- SOAP over HTTPS,
- A proprietary binary protocol over standard TCP/IP.

The principal advantage of the binary protocol is that it is more compact and thus better suited to low-bandwidth conditions; however, HTTP can more easily traverse firewalls. The AJAX interface to the server was developed at the end of 2007, too late for the examples in the following pages; it has the advantage of being independent of the operating system platform. A conferencing session can support any combination of the client protocols, depending on the rules imposed by system administrators.

The primary job of the GeoConference server is to support the conference sessions, including the final sandwiching of the background map layers coming from the data connectors. Any given geoconferencing session is quite lightweight and even a single-processor server can support many active sessions concurrently. We assemble the background map at the conference server to assure that every participant in a given session sees exactly the same rendering of the map at the same time, two elements we consider essential to informed group decision-making.

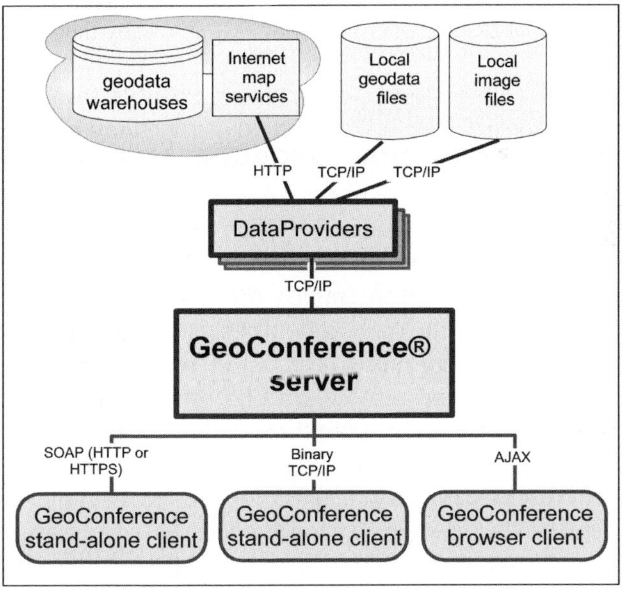

Fig. 3. Architecture of the GeoConference® system.

The GeoConference server relays user pointers, maintains the list of shared georeferenced annotations, and saves the last state of the conference session for later reuse. The saved session data includes the text-message log, the georeferenced annotations and references to (but not the content of) the layers of the map background and the map extent. The server maintains a small database containing user and workgroup information, saved sessions and data source information. The ability to fully record and replay conference activity will be added in 2008-09.

DataProviders are connectors to geoinformation sources; they act as dedicated mini-servers to the GeoConference server to which they are connected using standard Internet protocol. They work in two ways: as loosely-coupled clients to Internet map services (primarily WMS), or as connectors to file-oriented geodata and georeferenced imagery. They can also connect to non-georeferenced image files.

In the case of file-oriented data the DataProviders must also render the geoinformation layers for the conference server. Control for this rendering is passed through the conference server to the clients; for example, in the case of multi-band imagery, a remote sensing specialist can control band combinations, enhancement and simple filtering from within a GeoConference client. Data Providers run at the location of data source files and thus are under the control of data owners or brokers. In the case of connections to Internet map services, DataProviders may be at any Internet location, although efficiency suggests that they are best located close to either the map service or the GeoConference server.

12.2 Benefits of Real-time Geocollaboration

An all-too-common problem for EMOs across North America is the pressure to do more with less: better services from smaller budgets and staffs. The SCQ's answer has been to employ off-the-shelf Internet collaboration tools, one of which is GeoConference. It is used by regular SCQ staff, decision-makers and external partners, most of whom are not geomatics specialists. However, GIS and remote sensing specialists also use the system to introduce and discuss their analytical outputs with end-users and to receive ground-truth information from observers.

Participants in GeoConference sessions use lightweight, easy-to-deploy client software to take part in teleconferences based on synchronized map views. Using this tool, they are, in effect, performing "geocollaboration[1]", in the form of live, map-based, Internet conferences (that is, "geoconferences",

[1] *Geocollaboration* is a neologism formed from "geoinformation collaboration". While we are not sure who coined this term, we first saw it in material from the GeoVISTA Center in the Department of Geography at the Pennsylvania State University.

or "live geocollaboration"). The emphasis on direct user-to-user exchanges distinguishes GeoConference from other commercial software used for geocollaboration, which (to our knowledge, at least) are based on individual interactions with a shared database. In fact, we would expect these two types of geocollaboration to be complementary.

A secondary use of GeoConference at Québec Civil Security is as a georeferenced situation-monitoring tool. Typically, a geocollaboration session remains open in an operations center to allow it to follow events or general situations, (for example, the current locations and conditions of forest fires in the province). Users log in from time to time to add georeferenced observations (annotations) to the shared map or to check the current situation.

The user-controlled views in geoconferencing sessions are constructed of layers from multiple, distributed geodata and image sources. Teleconference participants can perform many of the same actions they might want to perform if they were physically present at the meeting – they can point at the map (multiple active pointers), draw on (annotate) the map, change the map extent, modify annotations or symbology, and add or remove layers. Typically, the map-based meeting accompanies a telephone conference attended by some or all of the participants. The system is designed so that foreground actions (pointer movements and shared annotation) keep up with the spoken conversation, even when some participants are using low bandwidth connections such as wireless or dial-up telephony. A logged, time-stamped text chat is also available; it can be used for instant messaging and as part of decision recording. The annotated map, which is stored as part of the session, also records elements of decisions.

Using the map-based, real-time collaboration tool has an important impact on SCQ costs and the services it provides:

- SCQ staff in the central or regional offices can be applied to problems in any region;
- Personnel from collaborating agencies can attend meetings without having to spend time travelling;
- Local officials can receive real-time guidance in the application of hazard and risk analyses;
- Outside experts can deliver and discuss their analyses and products without having to leave their offices or labs.

A collateral savings comes from the use of distributed geoinformation. The SCQ does not host or maintain all of the map information being viewed. Most of it comes instead from Internet sources such as OGC-standard Web Map Servers (WMS) or from GeoConference connectors at the data source. In addition to requiring less effort by SCQ staff and reducing duplication, access near the source provides access to fresher geoinformation.

During emergency operations (the response phase of EM), the use of an Internet-based, real-time geocollaboration tool virtualizes the maps in the operations center, which provides time savings and service improvements:

- Observers and operations personnel in the field can share the operational picture with the emergency operations center and with analytical information providers;
- Important geoanalytical products can be applied, discussed and modified without delay;
- Decision-makers can be fully briefed wherever they are located.

Using the examples in the following pages, we will try to illustrate some of the points mentioned above. The examples come from real cases or scenarios developed at Québec Civil Security. Note that the authors have translated some of the annotations and texts into English and in some cases recreated scenarios in order to capture screens for the illustrations.

12.3 A geocollaborative view of natural hazard and risk

Coastal erosion in the St. Lawrence River estuary seems to be accelerating, perhaps due to rising sea levels and reduced protection by shore-fast ice. This erosion can provoke landslides, debris flows and slumps that endanger lives, property and infrastructure. Geotechnical specialists in the Québec Ministry of Transport (MTQ) and the DGSCSI use slope, geomorphologic and soil information, along with orthophotos (for feature reconnaissance) and field observation to delimit the presence and likely effects of related geomorphologic processes (Fig. 4).

Recently, MTQ and DGSCSI personnel have been able to discuss their work in ad-hoc telephone calls or in formal teleconferences using the environment provided by the SCQ's real-time geocollaboration system, as shown in the figure. The ability to discuss geographic information at a distance avoids the disruption and cost of travel (even if this is just across the city) and so encourages more frequent exchanges. The geocollaboration software provides the ability to view everyone's information in the same geographic context. Since each participant controls a map pointer and can make and edit both the symbology and the georeferenced geometry of point, line or polygon annotations, discussion is precise and open.

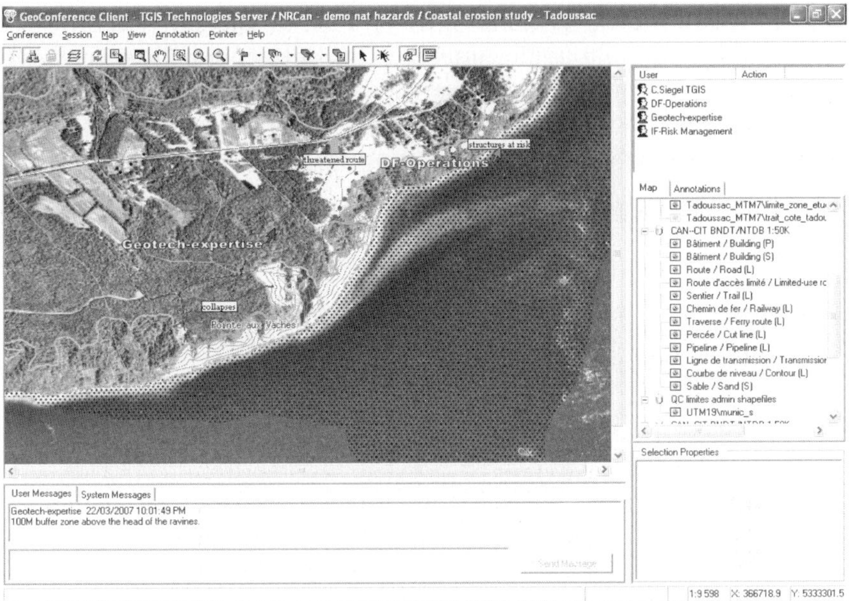

Fig. 4. Identifying costal erosion/landslide hazards based on reconnaissance using an orthophoto image, slope and soil information, and ground observation. Data visualized here comes from the Québec MRNF and MSP as well as from Natural Resources Canada. The screen capture and the following examples show the GeoConference® client being used for live map-based collaboration over the Internet.

12.4 Geocollaboration for communicating risk information

Once hazards are recognized, we need to see them in the context of land use and development. In Québec, the Ministry of Transport produces analytical cartography showing zones of risk for mudslides and mudflows (Bilodeau et al., 2005). Municipal governments are responsible for emergency management planning and land-use controls based on these risks. As part of its mission of risk reduction, Québec Civil Security supports and advises municipalities in the planning process.

Using the real-time geocollaboration system, SCQ analysts can help municipal officials put risk evaluation into local context, aid them in interpreting the risk zones and discuss development scenarios. In the example shown in Fig. 5, the municipality is several hours by route from the central office of the SCQ and an hour or so from the closest regional office. The ability to hold a teleconference, in full geographic context, with municipal officials means a large savings in time (and in travel expenses, too). Perhaps more to the point, it allows the SCQ to offer more frequent and better quality services to local government.

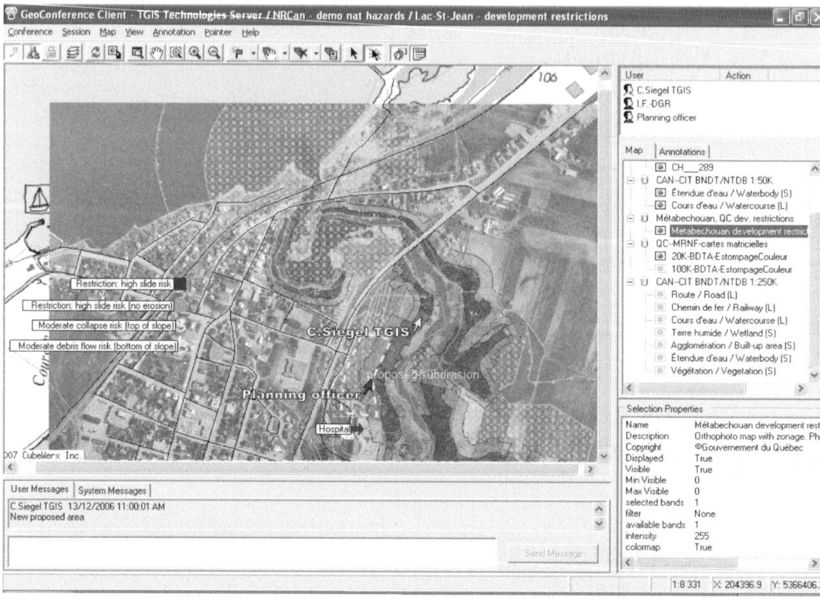

Fig. 5. Orthoimage with overlays indicating mudslide and mudflow risk zones. The orthoimage, along with information from the MRNF's 1:20 000 series mapping, serves as a part of a detailed georeferenced backdrop. Together with additional layers of information from Canadian and Québec WMS sources, they provide an understandable context for real-time discussions between personnel from the SCQ and local officials.

12.5 Real-time geocollaboration for emergency operations

One of the most obvious applications of real-time geocollaboration in emergency management is coordination and response. The advantages of doing so are similar to those mentioned in previous cases: savings in of personnel time and travel costs. However, beyond the simple question of cost, map-based teleconferences can deliver geospatial knowledge and expertise more effectively and swiftly – they are a service improvement. In a rapidly changing situation, collaboration at a distance may be the only way the information *can* be provided. The following example illustrates how map-based teleconferencing helps to communicate geospatial knowledge during emergency situations.

In winter and spring, river ice conditions (ice dams, frazil ice build-up) can provoke flooding. The SCQ monitors rivers, assists municipalities in the case of inundation, and, when necessary, coordinates action to reduce the risk or effect of flooding.

The Laboratory of Remote Sensing in the *Centre Eau, Terre et Environnement* of the *Institut national de recherche scientifique* (INRS-ETE) in Québec City has, over the past several years, developed techniques and procedures to rapidly produce synoptic, classified image-based maps of river ice from Radar satellite imagery (Fig. 6). The initial version of these river ice maps can now be delivered within a few hours of image acquisition (Gauthier et al., 2006).

In order for the SCQ to make effective use of these image maps, several problems have to be solved:

- The ice map has to be made accessible, rapidly and in company with the operational context (which includes information from many different sources);
- The image analyst needs to participate in the interpretation of the ice information for operational use;
- It is quite likely that image classification will have to be adjusted based on ground observations.

In the situation shown in the figure below, the image analyst was in Québec City and the flood situation in the Montréal region, more than 200 km distant. How would the analyst see and understand the current situation, discuss image interpretation and related ground observations in real time, and then provide an updated classification before the image information was out of date?

The real-time geocollaboration system provides ways of solving these problems (Siegel et al., 2004). Using a GeoConference connector for GeoTIFF format, which acts as a dedicated image server, the image analyst can provide the image to the geocollaboration system from within his own network. The GeoTIFF image file need never leave its original location. The analyst himself remains in his lab while explaining interpretation and getting up-to-date information from on-site observers that will allow him to improve image classification, which is in turn available in the geoconference. This activity occurs in the georeferenced context of all of the other information on the SCQ's operational map.

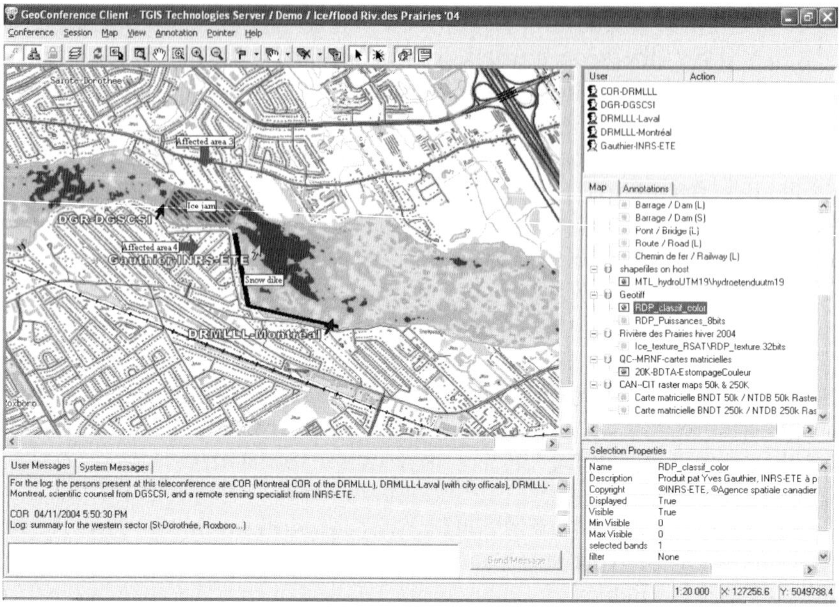

Fig. 6. Radar-based river ice map as part of the operational picture for EM during ice-related flooding in the Montréal region in late January and early-February 2004. Québec and Canadian federal map layers (WMS and vector file sources) provide part of the situational context. Georeferenced annotations on the map show information about current conditions and operations.

12.6 Forest fire situation monitoring and coordination

Wildfires become a civil emergency management problem when they threaten settlements, transport infrastructures, or economic exploitation activities. In Québec, a crown corporation, the *Société de protection des forêts contre le feux* (SOPFEU) has the responsibility for fighting and monitoring the fires themselves while measures to protect the forest resource are the responsibility of the *Ministère des Ressources naturelles et de la faune* (MRNF). Coordination of provincial efforts to protect the civil population, property and infrastructure falls to the SCQ.

Forest Fire Pilot Project. In the spring of 2006, Québec Civil Security initiated a pilot project using GeoConference for civil emergency management related to wildfires. Until that time, the SCQ's use of real-time geocollaboration had been sporadic and internal; that is, for coordination within the Ministry of Public Security. The objective of the pilot project was to roll out and use the system comprehensively in the primary forestry production areas. It was to be deployed to the SCQ's regional offices and to the other members of the regional emergency management organizations (ORSCs – see section 1 above), including the SOPFEU and regional representatives of the MRNF, as well as representatives of many other provincial bodies, and para- and non-governmental organizations such as the Canadian Red Cross. Teleconferences based on geospatial information were to become part of the formal coordination meetings of the ORSCs during emergencies, in effect 'virtualizing' the regional operations centers.

The MRNF and SOPFEU were geoinformation providers for the project as well as being participants. Natural Resources Québec (MRNF) provided access to WMS sources (hosted by CubeWerx, Inc.) showing base map information at 1:20,000, 1:100,000 and 1:250,000 nominal scales, along with a small-scale digital elevation model, forestry roads and recreational leaseholds on provincial lands. The *Centre de mesures d'urgence* of the MRNF provided two or more processed NOAA AVHRR images daily to aid fire and burned area detection. Provincial base mapping was complemented by a number of layers from the Canadian federal National Topographic Data Base (NTDB), which provided place names, transportation infrastructure information, and information about a number of risk generators. We accessed the NTDB layers through the Canadian Centre for Topographic Information's WMS.

The SOPFEU provided essential fire monitoring information: fire location, control situation (new, out-of-control, contained, controlled, extinguished), and fire contours. Since the SOPFEU did not have the facilities available to provide a geoinformation service, the data was transferred to the MSP using an automated file transfer protocol (FTP) process.

The 2006 fire season in Québec was a very quiet one, with the number and size of forest fires well below average. This situation effectively precluded use of the geocollaboration system for regional coordination meetings. The project was extended to the 2007 fire season, which turned out to be much more active, and the geocollaboration system went through several periods of intensive use.

Extension of the project for the 2007 season also provided an opportunity for the Québec Ministry of Public Security to develop its own Web Map Service to publish the SOPFEU's fire information for internal use. The data were harvested and the WMS layers updated every 30 minutes, following the SOPFEU's update schedule (fire contours, however, were generally updated at the beginning and end of the fire-fighting day). The figures that follow all include layers from the MSP's WMS.

Geocollaboration in the Pilot Project, 2007. During the fire season (May-September), the SCQ's provincial emergency operations center maintained a continuous daily, up-to-the-minute monitoring session using

GeoConference. This session was available for viewing and annotation by the regional operations centers or any other project participant (Fig. 7). We initiated sessions each morning in less than a minute using a session template.

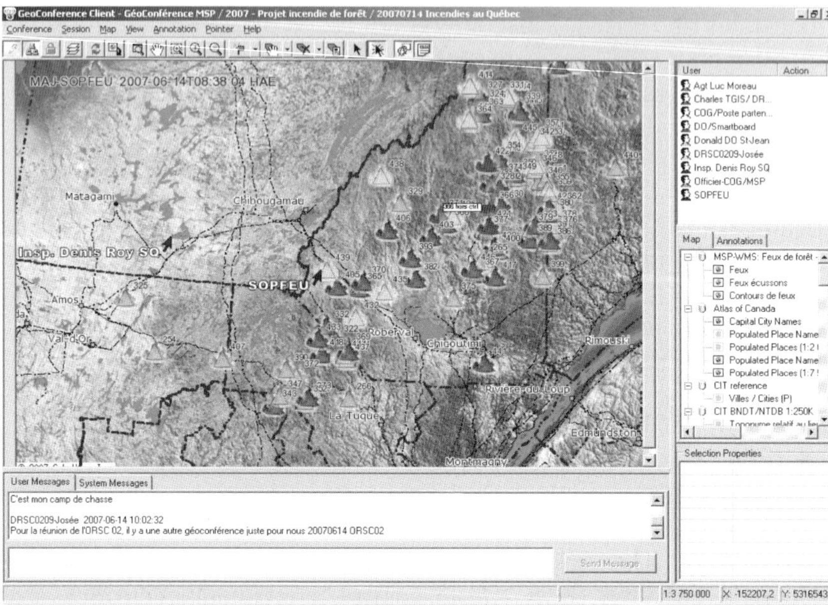

Fig. 7. Fire situation session, 14 June 2007, fire update from 8:38 AM. Different nuances gray of the fire symbols indicate the state of each fire (e.g. dark grey is 'controlled'). The time stamp in the upper left indicates the time of the most recent update received from the SOPFEU.

Additional geoconferences focused on particular regions or fires to supplement the daily provincial monitoring sessions; other conferences were set up for coordination meetings. More than 190 GeoConference sessions were saved over the 2007 fire season. Fifteen of these were used during formal regional EM coordination meetings (Fig. 8); often, they were used for both a morning and an afternoon meeting in the same day.

Frequently, partners relayed data files of map information in the period just before a coordination meeting. These data were immediately made available in the GeoConference sessions. Occasionally, project participants transmitted non-georeferenced images of maps for use during coordination meetings. Meeting participants viewed these in non-georeferenced sessions, which were accessed in parallel to the georeferenced ones (that is, users could view, discuss and annotate them in another GeoConference client window).

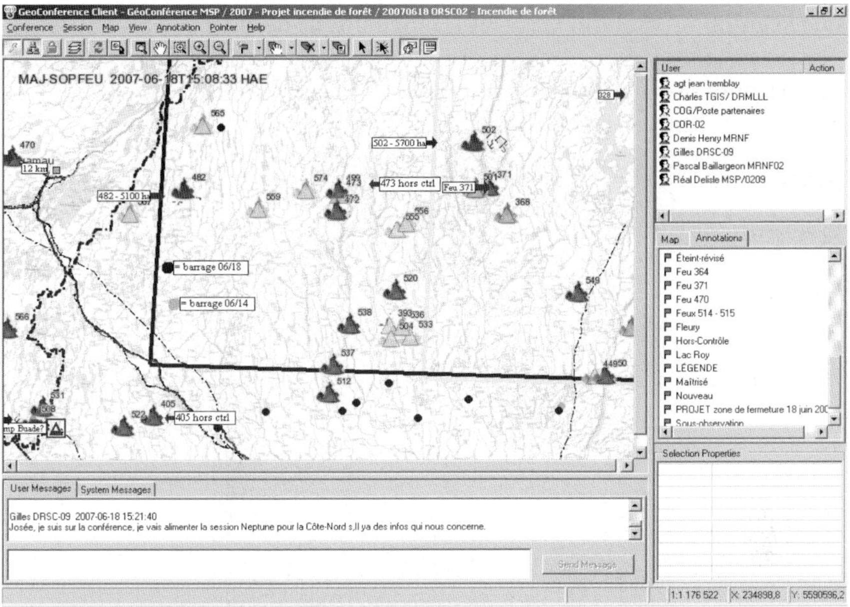

Fig. 8. Afternoon coordination meeting, 18 June 2007, ORSC region 2 (Saguenay–Lac-St-Jean Region), with fire information updated at 3:08 PM. The rectangle (partly visible) shows the zones closed to forestry activities on the 18th of June. The circles ('barrage 06/18' and 'barrage 06/14') represent roadblocks for two dates. Regional offices of the MRNF communicated the files containing the closure information directly to the SCQ staff.

During the two fire seasons of the pilot project, system pilots made GeoConference training sessions available to end-users (these sessions were in a separate workgroup and were not part of the 190-plus working sessions mentioned above). One hour-long initiation to the conferencing client seemed to be sufficient to make users competent to use the software's primary conferencing features (text messaging, map pointers and georeferenced annotations). Approximately five SCQ and 15 partner users were trained in each fire season. These new users frequently trained others in their work locations or regions.

Conclusions from the Pilot Project. Widespread participation of partners outside the MSP in geocollaboration sessions made it possible to follow rapidly changing fire situations (Fig. 9). Use of the system provided a channel for updating geoinformation, augmenting both the speed and the precision of situation evaluation. Given the distances involved, many participants in coordination meetings could not attend in person and Internet map conferencing became an essential addition to telephone conferences.

Users seemed to find the GeoConference client easy to use and appreciated the possibilities for situation monitoring and attending coordination meeting without leaving their work locations. They were able to manage or participate in online, map-based meetings with very little training.

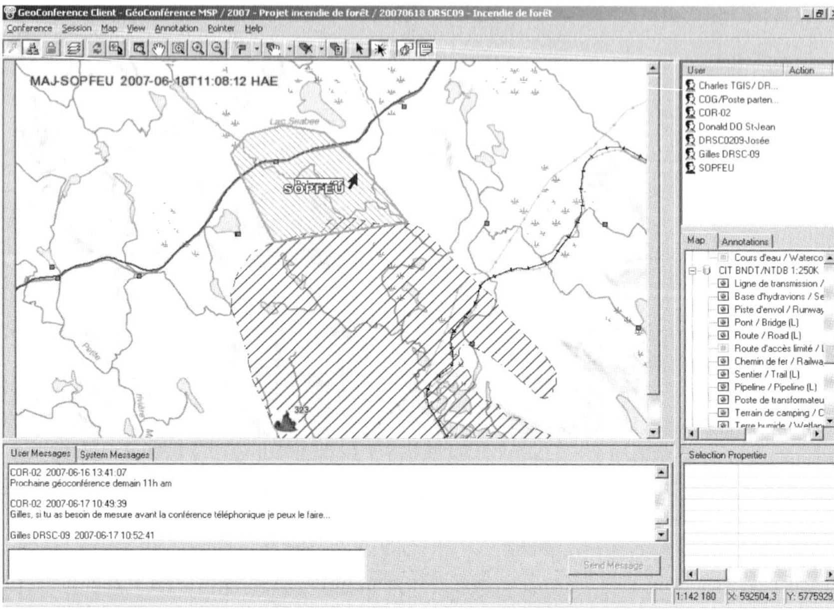

Fig. 9. Internet meeting for ORSC region 9 (North Coast), 18 June 2007. The smaller polygon above updates the burned area for fire 323, which had just gone out of control. The SOPFEU generally publishes fire contours only one or two times daily, so the ability to manually draw an update is crucial when fire growth threatens infrastructure, settlements or camps. [Apologies from the primary author: I had scrolled up to look at previous messages so the screen capture does not show the relevant part of the text log].

Staff at the provincial operations center in Québec City and other specialists in Montréal and St-Jean-sur-Richelieu supported coordination meetings for regional operations centers. This left regional counselors free to run meetings or assist each other. The collaboration system lets everyone profit from the capabilities of the geomatics and thematic specialists.

The pilot project showed once again that distributed sources, delivered through a Web services infrastructure, can easily supply much of the geoinformation required for emergency management. However, EM does require at least a limited amount of base map information on a guaranteed, 24/7 basis, so a copy of at least some base geoinformation is likely to be maintained by the SCQ itself. The project also demonstrated that Web services can satisfactorily supply up-to-the-minute, rapidly changing thematic information. We should note, though, that in the case of one-off situational information, such as road closures, we used file-oriented data delivered directly from partners.

12.7 Conclusion

Emergency management, in all phases, is an activity requiring the cooperation of multiple organizations and specialists. Geospatial analysts need to take an active part in the collaboration. Frequently, this collaboration will take place at a distance, since teleconferencing is a standard tool of the EM. A live collaborative view of geoinformation must become an integral part of the teleconference. Another characteristic of the solution is rapidly distributed, easy to use, and in-context geoinformation for all users, not just geomatics specialists. A real-time Internet geocollaboration tool that is usable by the non-specialist provides improvements in service and efficiencies, in terms of both cost and time, to the emergency management organization. It also adds value and immediacy to the products of the geospatial analyst.

The larger conclusion of the SCQ's experience with live geocollaboration is that it will extend Internet map conferencing to full production use for all regions and for all risks. The organization is also hoping to increase the integration of its real-time Internet system for incident management and map conferencing.

The synergistic value of geocollaboration with the SCQ's partners is evident. On that basis, the organizations has been asked to prepare a business plan for deployment of the live geocollaboration tool to all Québec departments and organizations with responsibilities in civil defense

Acknowledgements

River ice mapping in Figs. 2 and 5 is ©Institut national de recherche scientifique, Centre Eau, Terre et Environnement. It is based on Radarsat image data ©Canadian Space Agency.

Some of the cartographic layers in Figs. 2–8 are from the *Ministère des Ressources naturelles et de la Faune du Québec* and the *Ministère des Transports du Québec,* ©Government of Québec. Thanks to CubeWerx inc. of Gatineau, Québec for hosting some of this geoinformation and making it visible through a demonstration WMS using their CubeServ software.

Some cartographic layers in Figs. 2–9 are from the Centre for Topographic Information, Department of Natural Resources (Canada), ©Her Majesty the Queen in Right of Canada (furnished through the Centre's WMS).

GeoConference® screen captures in Figs. 2–5 in this paper have been taken from sessions hosted at Natural Resources Canada's Innovation Acceleration Centre (part of the Earth Sciences Sector). Screen captures in Figs. 6, 7, and 8 are from sessions hosted by the Québec Ministry of Public Security.

GeoConference is a registered trademark owned by TGIS Technologies inc. of Chelsea, Québec, Canada. TGIS Technologies developed the initial version of GeoConference software in a project in the GeoInnovations Program in the GeoConnections Secretariat, Natural Resources Canada. The DGSCSI of the *Ministère de la Sécurité publique du Québec* acted as one of the primary end-user-collaborators in the project. The development of the GeoTIFF connector for GeoConference was partially supported by the Canadian Space Agency.

References

Bilodeau C, Genois A., Demers D, Potvin J, Robitaille D, Dupuis C, (2005). Cartographie des zones exposées aux glissements de terrain dans les dépôts meubles : Saguenay - Lac-Saint-Jean : guide d'utilisation des cartes de zones de contraintes et d'application du cadre normatif. Ministères de la Sécurité publique, des Transports et des Affaires municipales et des Régions.

Committee on Intersections Between Geospatial Information and Information Technology (2003). IT Roadmap to a Geospatial Future. Report prepared for the Computer Science and Telecommunications Board, Division on Engineering and Physical Sciences, National Research Council of the National Academies (USA).

Gauthier Y, Weber F, Savary S, Jasek M, Paquet L-M, Bernier M (2006). A combined classification scheme to characterise river ice from SAR data. EARSeL eProceedings, 5(1): 77-88

Siegel C, Pellerin É, Fortin D (2005). Bringing Geospatial Expertise to Emergency Operations Management with Geoconferencing in *Proceedings of the 98th Conference* of the Canadian Institute of Geomatics, June 13-15, 2005, Ottawa, Ontario, Canada.

Siegel C, Gauthier Y, Fortin D (2004) La GéoConférence: un outil de communication innovateur pour l 'intégration de la télédétection dans la gestion des situations d'urgence. Bulletin de l'Association québecoise de télédétection.

Acronyms

EM	emergency management
EMO	emergency management organization
EOC	emergency operations center
INRS-ETE	Institut national de recherche scientifique, Centre Eau, Terre et Environnement
DGSCSI	Direction générale de la sécurité civile et de la sécurité incendie, MSP (Québec)
MRNF	Ministère des Ressources naturelles et de la Faune du Québec
MSP	Ministère de la Sécurité publique du Québec
MTQ	Ministère des Transports du Québec
NRCan	Natural Resources Canada (Department of Natural Resources, Canada)
OGC	Open Geospatial Consortium
RS	remote sensing
SCQ	Sécurité civile du Québec (also referred to here as Québec Civil Security), DGSCSI, MSP (Québec)
SOAP	Simple object access protocol (currently only the acronym is used)
SOPFEU	La Société de protection des forêts contre le feu
WMS	Web map service (an OGC-defined interoperability standard)

13. On-line Street Network Analysis for Flood Evacuation Planning

Darka Mioc, François Anton and Gengsheng Liang

Abstract

Flooding has been one of the most costly disasters in terms of property damage and human casualties. The damage to the infrastructure can also be extensive, and it can endanger flood evacuation efforts. During flooding, emergency managers need to know which evacuation route is safe for their vehicles. In evacuation planning, knowing the water depths on roads and taking into account vehicle types available for rescue is highly demanded. This paper presents a Web GIS system for dynamic street network analysis in evacuation planning, which includes a flood monitoring and prediction system, a street network and flooding spatial data model, and an application for analysis of street network flooding in near-real time conditions. The results obtained in this research demonstrate that users can obtain evacuation routes in different situations, where water depths and vehicle types are considered as constraints for the street network. The use of this system enables emergency management services and the general public to make a timely and effective decision for flood evacuation planning.

13.1 Introduction

Floods are common natural disasters that cause much damage to people's lives and properties. Due to population growth and the accompanying development of extensive infrastructures close to rivers, flooding disasters impacts have greatly increased over recent years. An effective GIS tools for evacuation planning could help mitigate the worst effects of flood disasters.

Experience with emergency evacuation in the Lower Saint John River during the flood of 1973 demonstrated the problems people encountered (Environment Canada, 1974). At the peak of the flood in spring 1973, basements and ground floors of buildings near the river were flooded (Fig. 1), and roads and bridges were damaged as well. In these conditions, the management and control of the traffic and the flood evacuation became a critical problem (Fig. 2).

Fig. 1. Flooded Houses (From http://www.gnb.ca/public/riverwatch/Flood-1973/index.htm)

The evacuees rescued from their homes by boats or amphibious crafts (Fig. 3) needed transportation to reception stations.

Fig. 2. Flooded road (From http://www.gnb.ca/public/riverwatch/Flood-1973/index.htm)

Fig. 3. Flood rescue with amphibious craft(From http://www.gnb.ca/public/riverwatch/Flood-1973/index.htm)

The transportation routes needed to be redesigned due to isolation and flooded highways (Fig. 4).

Fig. 4. Road closure caused by flooding (From http://www.gnb.ca/public/riverwatch/Flood-1973/index.htm)

In reality, many people know where the refuge shelters are, and how to find the way to escape the flood. Nevertheless, under flooding, the people might not know whether their usual path is safe or not since some roads may be inundated. It could be dangerous for people to rescue their lives and properties through flooded roads. In some flood rescue cases, emergency teams or individuals try to drive their vehicles through flooded streets if it seems that it is shallow water (Fig. 5).

Normally, people should not drive through flooded roads. In some circumstances, however, individuals and rescue managers may want to know water depths of inundated area in order to allow rescue vehicles to pass for quick rescue/evacuation through shallow water. Therefore, how to obtain the safe evacuation route considering dynamic information on water depth on roads and vehicle types becomes important for emergency managers.

Fig. 5. A bus trying to pass an inundated area (From http://www.gnb.ca/public/riverwatch/Flood-1973/index.htm)

Dynamic information on the flooding that affects transportation infrastructure is essential for flood evacuation planning in order to make the decisions for rescuing people's properties through transportation networks (Asama, 2006; Holz, 2006).

An effective flood monitoring network system that can offer dynamic information could help mitigate the worst impacts of flood disasters.

During flood events, the accessibility of the street network might change due to inundations of the streets or damages caused by flooding. Thus, the optimal path for flood rescue computed using GIS functionalities might not be valid if the system cannot integrate the flood information dynamically.

Geographic Information Systems (GIS) can play an important role in flood evacuation. Not only that they can integrate a variety of data for presentation in the form of digital maps, but also offer different spatial analysis tools needed for flood evacuation. At the same time, with the advance of Internet technologies, Web GIS provides an effective way to distribute geospatial information on the Internet, and makes the GIS data management and application accessible to emergency managers and the public.

In the following sections, we will present a Web GIS application for flood evacuation planning. The application displays flooded streets on-line and provides the evacuation path, which considers water depths of flooded areas and types of rescue vehicles. The application integrates a Web Mapping Server, a spatial database, and a network analysis application. A detailed implementation and results of the system will be presented using Fredericton's street network. This Web GIS application enables emergency managers to plan road closures and the accessibility of evacuation routes for different types of rescue vehicles.

13.2 Flood evacuation overview

Flood evacuation is the act of rescuing people from a dangerous place to a safe place during flood events. Once a flood event occurs, emergency services have to plan safe and short routes from refuge places to evacuation places. Lack of this knowledge would cause a loss of valuable time in the evacuation process (Simonovic and Ahmad, 2005). Emergency management services need to handle many evacuation processes, for example, they have to identify areas to be evacuated, sheltering needs and capabilities, and develop evacuation movement control procedures (Schwarzenegger and Renteria, 1997).

Numerous research efforts have been conducted in the past to develop new approaches for flood prediction and flood rescue using GIS. Several research efforts focused on dynamic flood modeling and prediction. Al-Sabhan et al. (2003) presented a real-time hydrological model for flood prediction using GIS. Zipf and Leiner (2004) in their research work proposed the scenario for a mobile flood warning system and gave an overview of system architecture they developed. Research in flood evacuation brought several different approaches so far. For example, (Lu et al., 2003) proposed two heuristic algorithms named Single-Route capacity constrained planner and Multi-route capacity constrained planner to incorporate the constraints of routes. This research focused on traffic difficulties such as congestion and accessibility. The novel research by Neis, (Neis et al., 2007) on 3D emergency route services is based on Open Location Services including a directory service, a gateway service, a location utility service, a presentation service, and a route service. Although the route service allowed various criteria settings such as time, start and destination points as well as the function to add avoid lists; the output routes could not allow shallow water on the roads. A genetic algorithm to retrieve the optimal shelter locations for the flood evacuation planning was presented by (Kongsomsaksakul et al., 2005). The algorithm considered the allocations of both authority and evacuees, but did not take water depth on roads into account. There were only few approaches considering the effect of water depth in flood evacuation. The approach presented by (Liu et al., 2006) proposes an adaptive evacuation route algorithm considering the effect of the change in water depth on people's evacuation behavior during flood events. However, efficient real-time data collecting, processing, and optimal evacuation route calculation should be given more concern in flood prediction and rescue.

The review of the recent research in flood evacuation shows that efficient methods of data collection and processing needed for safe evacuation paths planning which consider water depths on flooded streets and emergency vehicle types is still missing.

Three main problems regarding evacuation routes planning for flood rescue can be drawn from the review of current research on methods for flood evacuation:

- **Parameters affecting the optimal route are not enough**. Besides normal criteria such as start and end points, distance, street restrictions, water depths on streets, vehicle types and time should be considered.
- **The relationship between streets and flooded area is not clear**. Conventional network analysis can consider inundated streets as inactive parts of the network. In reality, the inundation depth levels should be seriously considered as criteria for planning the emergency evacuation.
- **Real-time data acquisition and communication are costly**. Many algorithms for flood evacuation are based on static data. However, since the flood water levels change continuously, it is important to allow real-time data being processed dynamically within GIS application.

13.3 Development of the flood monitoring network system

13.3.1 Real-time data accessibility

On-line GIS tools can help users to obtain near real-time information needed for flood evacuation planning. On-line accessible GIS can provide users with user-friendly interface, interactivity, and instant access to the data. Furthermore, the general public can access these tools at low cost. This direct access to GIS tools and data is a critical factor in flood evacuation planning.

The Saint John River forecast system implemented after the record high flood in 1973 is now operated by the New Brunswick Department of Environment's Hydrology Centre. In the past, although the flood monitoring system provided by NB Department of Environment had very good flood prediction capabilities on water levels in different hydrometric stations (MacLaren Atlantic Limited et al., 1979), the floodplain delineation was manual, tedious, and labour-intensive.

In order to improve the current flood prediction system for the Saint John River, a new project has been initiated. Several provincial organisations in New Brunswick (Emergency Measures Organisation, NB Department of Environment, River-Watch and the University of New Brunswick) have been actively involved in this new project entitled 'Decision Support for Flood Event Prediction and Monitoring (FEPM)'.

The main objective of this project is to build up a decision support system to improve the prevention, mitigation, response, and recovery from flood events.

The New Brunswick Department of Environment Hydrology Center monitors a wide range of information on factors affecting flooding such as snow conditions, temperatures, precipitation patterns, water levels and stream flow conditions by using a wide variety of telecommunication systems ranging from satellites to the telephone.

The New Brunswick Department of Environment Hydrology Center team uses hydrologic modelling software (DWOPER[1]) to predict water levels for the next 48 hours along the lower Saint John River Valley by inputting climate data, weather forecast data, snow data, and flow data from approximately 60 water level gauges in the New Brunswick portion of the river.

[1] DWOPER a one-dimensional routing model developed by the Hydraulic Research Laboratory of the United States National Weather Service (Fread, 1992).

To support near real-time flood modeling, we developed the procedures for transmitting real time water level data from the New Brunswick Department of Environment – River Watch to the end users (Fig. 6). The water level data from the output of flood modeling by the Hydrology Centre in the Department of Environment are transmitted via FTP. The timestamp of new data will be checked every 30 minutes for upload in the database. Then the water level data are transferred to the FEPM Web Page for generating and displaying gauge bar graphs.

At the same time, the water level data are accessed by the software module for flood plain computations (Mioc et al., 2007). With the advent of robust computer tools and high accuracy Digital Terrain Model (DTM), automated floodplain delineation is achievable (Noman et al., 2003). Recently, several management systems for floodplain delineation have been developed and applied in the flood event areas. These include floodplain delineation using watershed Modeling System (WMS) (EMRL, 1998), Arc/Info MIKE11 (DHI, 2004), and HEC-GeoRAS (Ackerman, 2005). All of the above systems are required to combine the output of the hydrological model with the ArcGIS system.

In this project we used CARIS GIS software to implement floodplain delineation. CARIS (Computer Aided Resource Information System) develops and supports rigorous, technologically advanced geomatics software. CARIS GIS is a computer-based system for managing spatial and non-spatial data. It includes data capture and input, manipulation, query, analysis, and visualization functions. CARIS software supports Triangulated Irregular Networks and implements several algorithms for Digital Terrain model, such as interpolating elevations. CARIS can be integrated with hydrologic modelling for generating the floodplain maps.

Now, with the advance of GIS tools and high accuracy Digital Terrain Model available with LiDAR data acquisition, floodplain delineation can be computed in near real time (Mioc et al., 2007). From the water levels we can create a water surface Triangulated Irregular Network (TIN). Then, we compare the ground surface elevations with the water surface TIN in order to obtain the floodplain depth datasets. Calculating the height difference between these datasets, we can obtain the floodplain delineation and the flood polygons can be calculated automatically. Based on the dynamically updated water depth datasets, new floodplain extent and contour files can be generated frequently. When the flood plain has been generated, it is displayed as a flood polygons map accessible on Internet (Fig. 7).

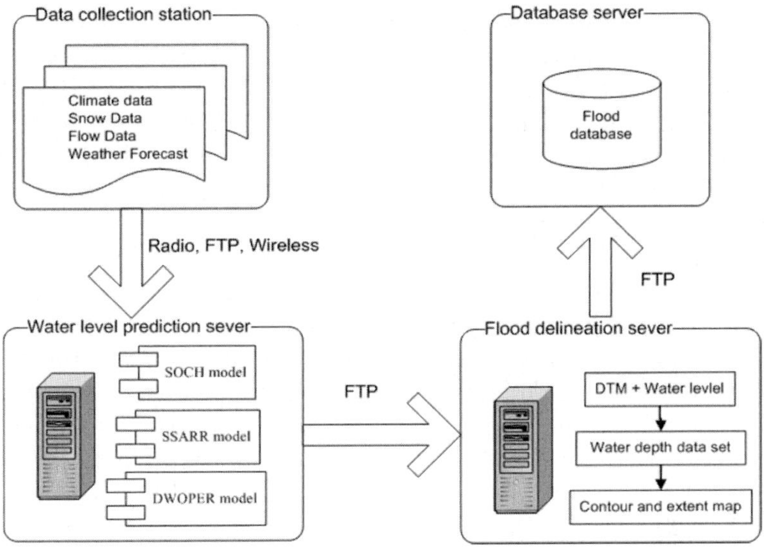

Fig. 6. Near-Real time hydrological modeling

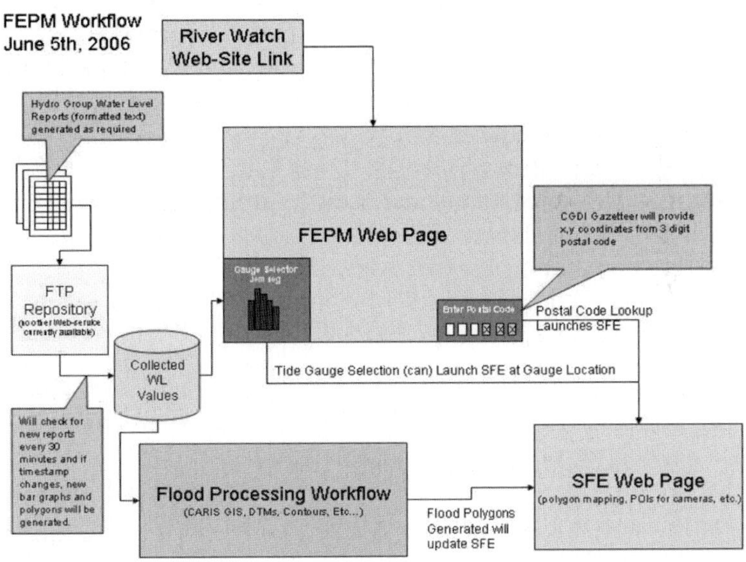

Fig. 7. FPEM workflow

Each time the new water levels are sent from the water gauges, the new flood polygons will be generated and accessible from the CARIS Spatial Fusion Enterprise Web Page (Fig. 7).

13.3.2 Data acquisition

The GIS data including buildings, roads, river, and flood data were prepared and imported in the system. These data were obtained from the City of Fredericton in an ESRI shape file format. The base map for this application is shown in Fig. 8.

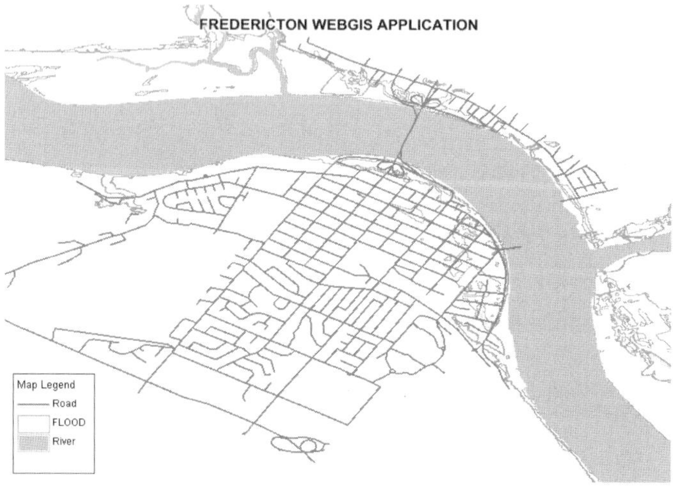

Fig. 8. The base map of Fredericton

The original data sets (streets and basic hydrographic features) were imported in an Oracle database.

13.3.3 3 Web GIS system for evacuation planning

The computer based flood monitoring and prediction system (Mioc et al., 2007) and wireless data transmission capability make available near real-time data needed for flood evacuation planning.

To display the flood information on streets and retrieve the evacuation routes via the Internet, a Web GIS system was designed. It utilizes a three-tier Web GIS architecture: Client Side, Server Side, and Back Side (Fig. 9). The browsers contact the Server side through HTTP in a request/response model. The Server Side includes a Web server and a spatial server with a network analysis application. This application enables users to update street network and process spatial analysis such as retrieving the shortest path in the network. The Server side is associated with the database through Java Database connectivity (JDBC).

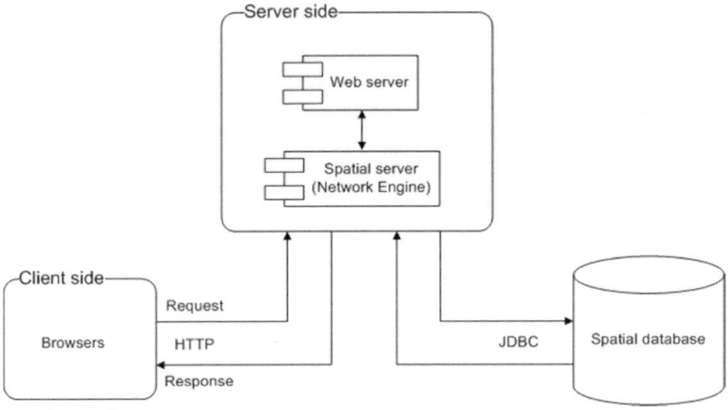

Fig. 9. Overview of system components

The system is composed of three components: a Web GIS (CARIS Spatial Fusion Enterprise), a flood monitoring and prediction system (Figs. 10 and 11), and an evacuation path analysis system.

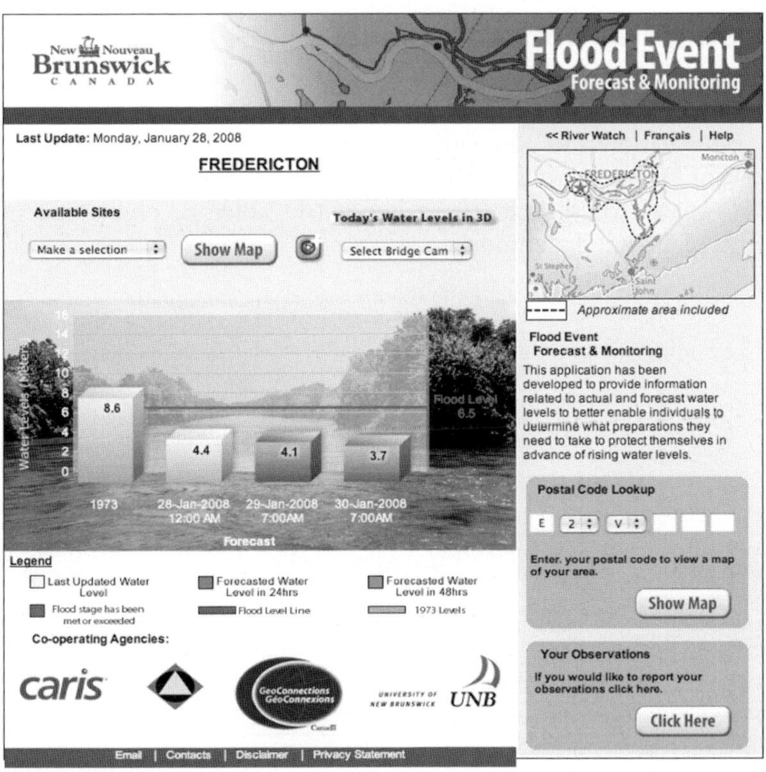

Fig. 10. Flood monitoring and prediction system

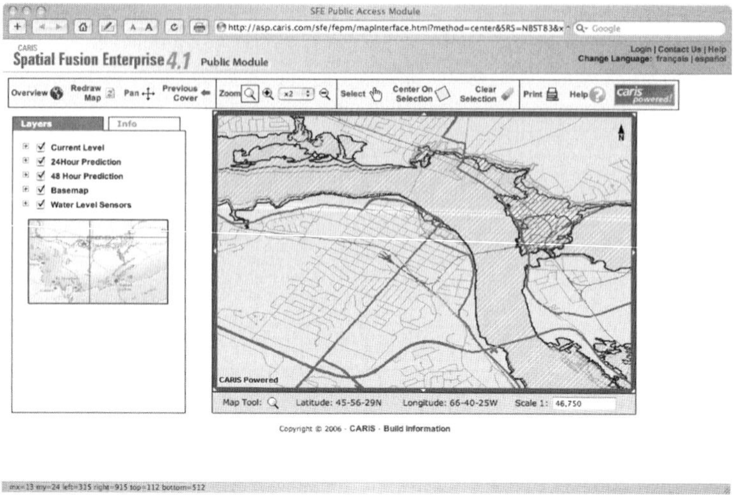

Fig. 11. Predicted flood maps available on-line

13.4 Street network data model

In this research, a street network data model that contains both connectivity information and geometric information was designed for flood rescue (Fig. 12). We have used the Oracle Spatial 'network with topology geometry' spatial data type (i.e. a geometric network embedded in a full topological spatial data model).

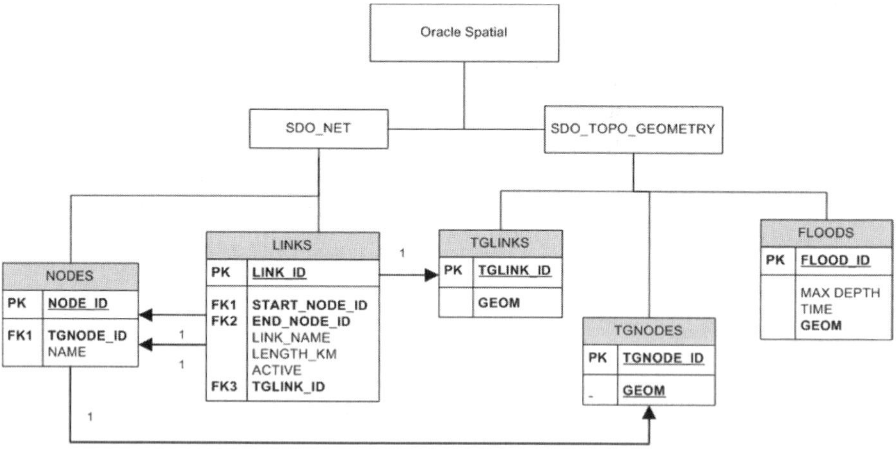

Fig. 12. Street network data model

The street network is stored in a SDO_NET logical network (nodes and links tables). The geometric embedding is stored in the Oracle topology geometry data type (i.e. a full topological spatial data model). Nodes represent intersections of streets, while links represent streets between two nodes. We can differentiate the logical nodes and links (with no geometric embedding) stored in the SDO_NET data type, from their embedding in a geometric space with full topology stored in the SDO_TOPO_GEOMETRY data type.

The geometric embedding of nodes and links and flood polygons are based on the Oracle topology geometry data type (SDO_TOPO_GEOMETRY), which is compatible with OGC simple feature access protocol. Considering the flooding events, the designed street network has a flood table that includes flood information on streets such as start point, end point, maximal water depth, and time. The flood polygons represented with the contour lines of different water depths allowed for rescue vehicles are accessible from the dynamically generated flood planes. The relationship between links and flood polygons is many to many.

13.5 Network analysis application

Based on the past experience with flood evacuation in New Brunswick during the flood in 1973, a prototype of network analysis application for flood rescue planning was designed in this research.

In GIS, since the dynamic flood information cannot be integrated automatically, and it is missing in the spatial database, the network analysis application only considers relatively simple variables: network connectivity, node and link state, and costs. For this project we developed additional applications: calculation of flood attributes and calculation of the shortest path for evacuation planning under flood constraints. The shortest path algorithm (available as one of the ORACLE functionalities[2]) was extended to include the constraints for this application and considers the water depths on streets and vehicle types available for the rescue operations.

13.5.1 Intersection of flood area with street network

For the network analysis application we need to input the datasets of flood areas. Using automatic floodplain delineation, the Web GIS application for Flood prediction and monitoring hosted by the New Brunswick

[2] The evacuation route computation is based on Dijkstra's shortest path algorithm.

Department of Environment can provide the water depth contours for flooded areas. When we populate these contours lines together with the geometric embedding of the street network, we can run an intersection query that will return all the streets that are in a flooded area. For example, the intersections between flood plain and street network will return all streets that are inundated and the depth dataset of the streets in the flooded area is generated. The maximal water depth is the maximal value of the Flood contour lines that intersects with street network.

13.5.2 Evacuation routes

Once we have implemented the intersection of flood area with the street network we can calculate the evacuation route considering inundations of the streets and different vehicle types.

Fig. 13. Fire truck used for emergency operations

For this application we assume that the flood level is not changing dramatically and that flooding does not damage the road foundation.

In our application, each link in the street network has an attribute, which is used to define the class of each water depth.

The attribute – level ranges from 1 to 5 (Table 1). The level 1 means there is no inundation on these streets. Any vehicle can pass through these streets. The level 2 shows the water depth up to 0.2 meters. Only all terrain vehicles can pass these areas. The level 3 is between 0.2 and 0.5 meters of water depth and only the Fire trucks can pass these areas in extreme emergency (Figs. 13 and 14). The level 4 is more than 0.5 meters up to 1 meter of water depth. Only amphibious crafts can drive through these areas (Fig. 3). The level 5 is more than 1 meter. Only amphibious crafts can drive through these areas and the rescue is considered dangerous.

Table 1: Maximal water depth on streets

Level	Water depth (m)
1	0
2	(0, 0.2)
3	(0.2, 0.5)
4	(0.5, 1.0)
5	more than 1

Table 2: Traffic class and size

Size	Traffic class
1	Car
2	All-terrain vehicle
3	Fire truck
4	Amphibious craft or boat
5	Amphibious craft or boat in dangerous areas

Fig. 14. Cars on the flooded road (From http://www.gnb.ca/public/riverwatch/Flood-1973/index.htm)

After defining the basic network data model and the network constraints, we can design the network analysis application. In our application, we need to provide different water depth allowance for the vehicles as the inputs and the street network will be re-computed for this type of vehicle.

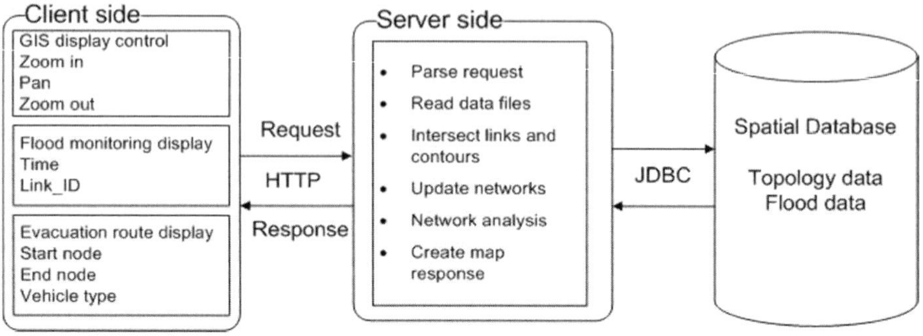

Fig. 15. Schematic diagram of procedures in the application

As shown in Figs. 9, 10 and 11, the Client side displays the flood situation map and helps users to compute evacuation routes by inputting vehicle type, time, and start and destination points. The user interface has a toolbar that is used for graphical display control, such as Zoom in, Zoom out, and Pan. In the Server side, the Web server parses the request to the spatial server. The spatial server accesses the flood and topology data from the database, processes spatial analysis such as intersection between street network and flood contour map, updates the street network with the current flood data, computes the shortest path in the constrained street network, generates a map, and then sends back to the Web server. Through the response of the Web server, flood maps with the evacuation routes will be shown in the client side. Street network, flood information, and other spatial information are all stored in the database (shown in Fig. 12).

13.6 Implementation of the prototype application

As shown in Fig. 15, the flow of application operation is as follows:

- The users construct a Web service request to obtain a map, and send request to *MapViewer* servlet through HTTP.
- The *MapViewer* servlet reads the related map definitions, fetches responding network which have been created in spatial database, proc-

esses spatial analyses such as getting the optimal path, and generates a map, construct an response.
- The users get the response and display it in the browser.

To create a user-friendly application interface, a number of JSP and HTML programs were developed. The HTML page allows users to select parameters such as traffic classes, time, and locations, click action buttons for zooming and panning, and view image map of the shortest path and flooded area.

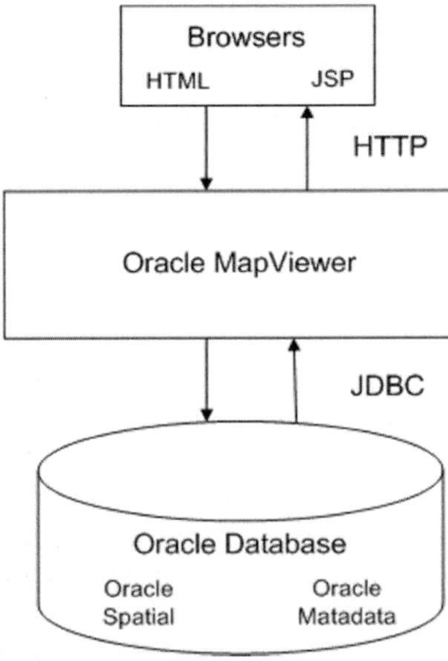

Fig. 16. System architecture

The whole program performs the following actions:
- Load the static and dynamic parameters. The static parameters include data source, the coordinates of the centroid, while dynamic parameters consist of network theme, map, and node label style, traffic size.
- Process the user actions, such as zoom in, zoom out, and submit.
- Construct a map request according to these parameters and the user actions.
- Send the request to the *MapViewer* servlet through an HTTP request.
- Parse the request, read the spatial data including network stored in Oracle database, update network, retrieve the shortest path, create the image map, and extract the image URL.
- Return the image URL to the HTML form.
- Create the user interface including map image, a user form, and several user action buttons.

As shown in Fig. 17, each map is associated with a set of themes, while each theme denotes spatial data from a specific table, and is associated with a rendering style. The application interface (shown in Fig. 17) allows that the flood information and network data can be viewed via on-line maps.

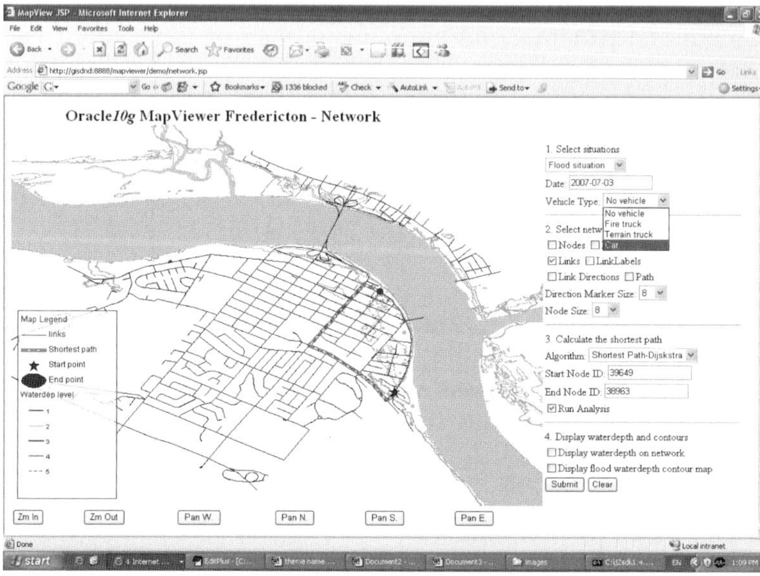

Fig. 17. The application interface

13.7 Results and Analysis

The results of this research provide the on-line accessible GIS tools for planning of evacuation routes and road closures. Using the flood monitoring network system, users can obtain the evacuation path information based on the water depth on streets and the type of rescue vehicles. Moreover, water depths on roads can be obtained and visualized in near real-time time.

As shown in Figs. 18 and 19, this application depicts water depths at different times during flood events. At the beginning of flooding, only a few streets are flooded (Fig. 18). At the peak of flooding, water depths on several streets have increased and the flood extent has expanded (Fig. 19). Thus, using the 24 and 48 hours flood prediction maps, users could obtain flood information on street networks in advance and have enough time to plan for evacuation.

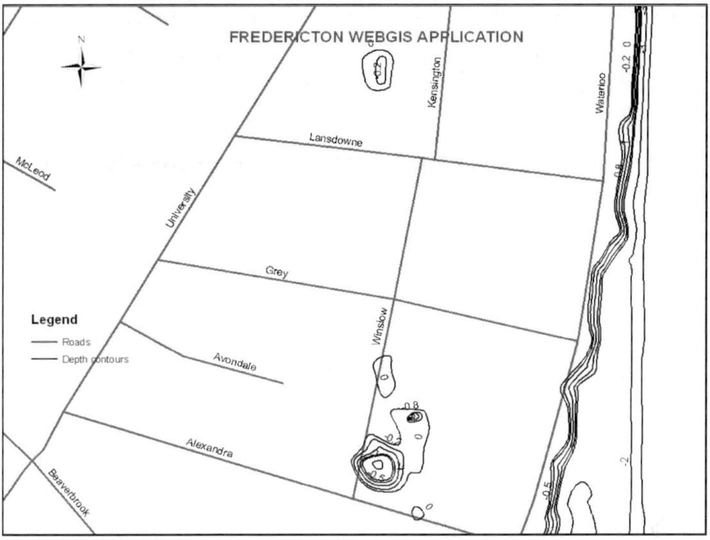

Fig. 18. Roads and water depth contours at the beginning of flooding

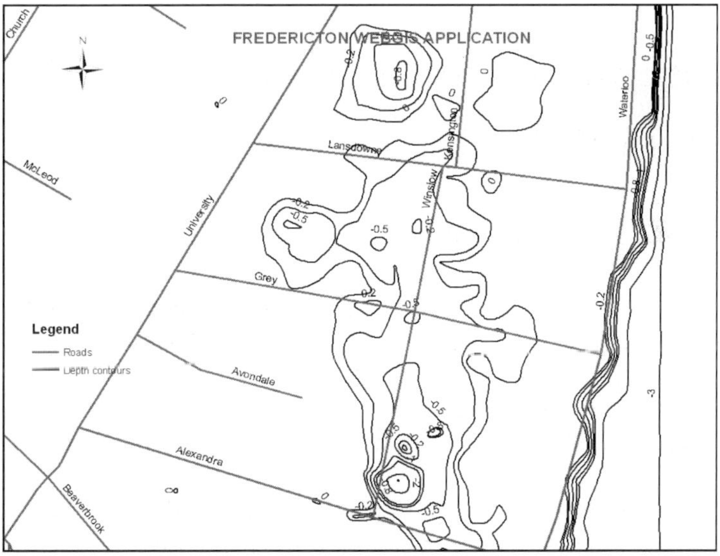

Fig. 19. Roads and water depth contours at the peak of flooding

A comparison of the evacuation routes for different types of emergency viechles was conducted to evaluate the application for flood evacuation planning.

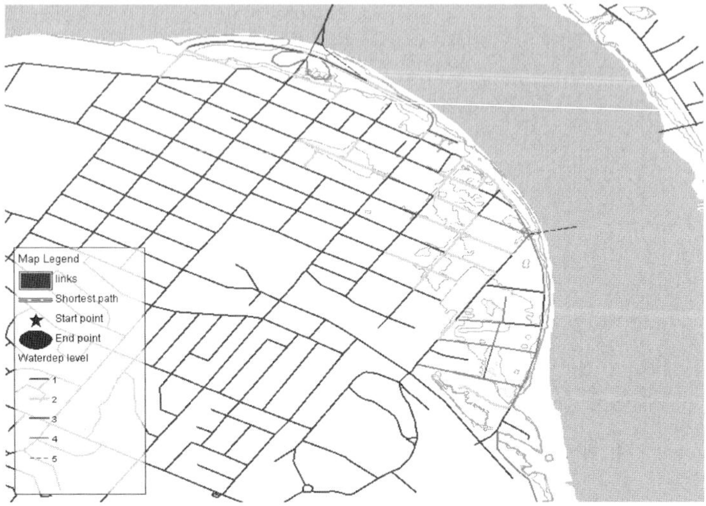

Fig. 20. Water depth on roads during peak of the flood in 2005.

Figure 20 shows the water depths on the roads calculated by intersecting flood maps and the street network. The streets that are inundated are labeled (shown in a distinct colour) according to the water depth predicted by the flood modeling application.

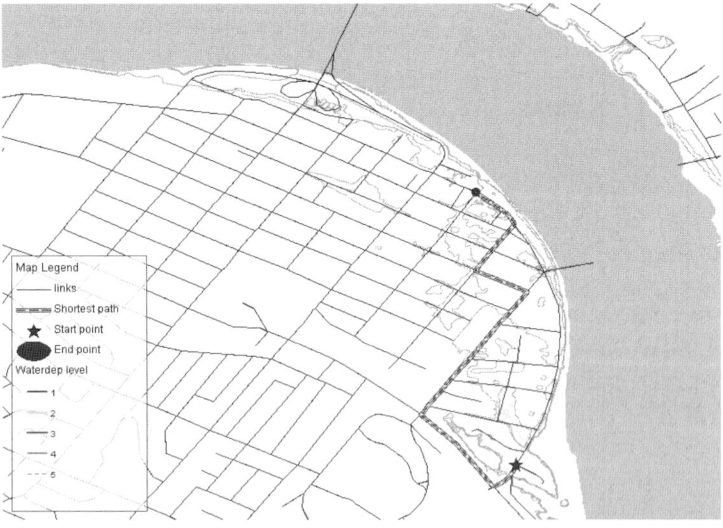

Fig. 21. The evacuation route for Fire Truck

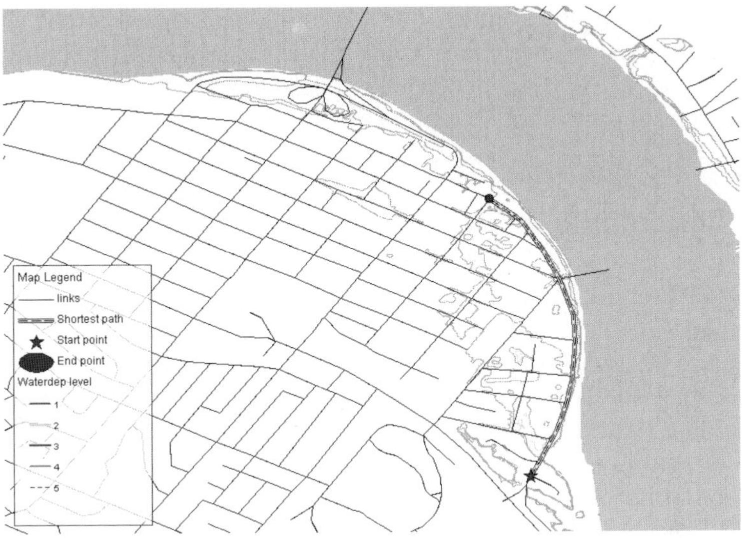

Fig. 22. The shortest path for amphibious craft

Figure 21 depicts another shortest path during the same flood event in 2005, but for another type of vehicle. Although some of the roads on its path are flooded, the water depth does not exceed level 3 (less than 50 centimeters) that is the water depth allowance for Fire Truck

Another calculation of evacuation routes was carried out to demonstrate the use of amphibious craft in the flood rescue operations. The map on Fig. 22 shows that the amphibious craft can pass even through the most inundated areas.

13.8 Conclusions

During flood events, the accessibility of the transportation networks could be affected by the raising water levels that can also cause damage to roads and bridges. The usual travel routes may not be safe due to the inundations of the roads. Emergency managers need to have near real-time information about the damages and inundations affecting the transportation infrastructure to effectively plan flood evacuation.

In this research, a flood monitoring network system has been developed as a GIS tool that can assist with emergency measures such as evacuation planning and road closures during flood events. It integrates a Web Server, a network analysis application, and a spatial database. The network analysis application takes the water depths on roads and emergency vehicle sizes into account to compute the shortest path for flood rescue. The Web GIS prototype implementation with Fredericton's street network, shows that the system can offer evacuation routes planning with different traffic vehicle sizes during flood events.

13.9 Limitations of this research

The uncertainty of the hydrological modeling and the accuracy of the elevation data acquired for DTM have to be considered as limitations in this research. The hydrological modeling for Saint John River can be accurate up to 20 centimeters and the accuracy of the digital terrain models is an additional constraint in this research.

In our future work, we plan to improve the hydrological modeling with the installation of additional water level sensors, and the accuracy of the Digital Terrain Model should be carefully checked in the areas susceptible to inundations and improved by GPS control points in the areas of interest.

Acknowledgment

This research work has received the financial support from the GeoConnections Secretariat of Natural Resources Canada for a project titled 'Decision Support for Flood Event Prediction and Monitoring'. The IT Division of the City of Fredericton kindly provided datasets available for this project. New Brunswick Department of Environment has provided data and expertise related to hydrological modeling and NB Emergency Measures Organization helped with their expertise and additional funding for this project. CARIS provided GIS software used in this project and CARIS contributed to the research project by providing the implementation for the part of this project within CARIS Spatial Fusion.

References

Ackerman C., T. (2005). 'HEC-GeoRAS GIS Tools for support of HEC-RAS using ArcGIS.' pp. 204.

Al-Sabhan, W., M. Mulligan, and G. A. Blackburn (2003). 'A real-time hydrological model for flood prediction using GIS and the WWW.' *Computers, Environment and Urban Systems,* 27(1), pp. 9-32.

Asama, H., Y., Hada, K., Kawabata, I., Noda, O., Takizawa, J., Meguro (2006). 'Rescue Infrastructure for Global Information Collection'. *Proceedings of the SICE-ICASE International Joint Conference,* Bexco, Busan, Korea, 18-21 October, pp.3443-3448.

Danish Hydraulic Institute (DHI), (2004) 'MIKE11 GIS reference and user manual.' 2004.

Environmental Modeling Research Laboratory (EMRL) (1998). 'Watershed modeling system (WMS) reference manual and tutorial'.

Environment Canada (1974). 'Flooding events in Canada-Atlantic Provinces.' [Online] December 02 2006. www.ec.gc.ca/water/en/manage/floodgen/e_atlan.htm.

Fread, D. L. (1992) 'Flow Routing.' Chapter 10, Handbook of Hydrology, Editor E.R. Maidment, pp. 10.1-10.36.

Holz, K. P., G. Hildebrandt, and L. Weber (2006). 'Concept for a Web-based Information System for Flood Management.' *Natural Hazards,* 38(1), pp. 121-140.

Kongsomsaksakul, S., Y. Chao, and C. Anthony (2005). 'SHELTER LOCATION-ALLOCATION MODEL FOR FLOOD EVACUATION PLANNING.' *Journal of the Eastern Asia Society for Transportation Studies,* 6 pp. 4237-4252.

Kothuri, R., E. Beinat, and A. Godfrind (2004). *Pro Oracle Spatial.* Springer-Verlag New York, New York, NY.

Liu, Y., M. Hatayama, and N. Okada (2006). 'Development of an Adaptive Evacuation Route Algorithm under Flood Disaster.' *Annuals of Disas. Prev. Res. Inst., Kyoto Univ.,* 49B.

Lu, Q., Y. Huang, and S. Shekhar (2003). 'Evacuation Planning: A Capacity Constrained Routing Approach.' *Proc. First NSF/NIJ Symp. Intelligence and Security Informatics,* pp. 111–125.

MacLaren Atlantic Limited, New Brunswick. Environment New Brunswick, and MacLaren Atlantic Ltd (MAL) (1979). *Canada-New Brunswick flood damage reduction program: hydrotechnical studies of the Saint John River from McKinley Ferry to Lower Jemseg.* Fredericton.

Mioc, D., L. Gengsheng, F. Anton, B. Nickerson (2007). Decision Support for Flood Event Prediction and Monitoring, *IGARSS 2007*, Barcelona, July 23-27, 2007, 4 pages, published on CD.

Murray, C. (2003). 'Oracle Spatial Topology and Network Data Models 10g Release 1 (10.1).'

Murray,C. (2004). '*Oracle Application Server MapViewer User's Guide 10g Release 2 (10.1.2).*' Oracle Corporation. [On-line] December 12 2006. http://download-east.oracle.com/docs/cd/B14099_01/web.1012/b14036.pdf

Neis, P., A. Schilling, and A. Zipf (2007). '3D Emergency Route Services based on OpenLS Specifications.' *Proceedings of Joint CIG/ISPRS Conference on Geomatics for Disaster and Risk Management,* Toronto, May 23-25, 2007.

Noman, N. S., E. J. Nelson, and A. K. Zundel (2003). 'Improved Process for Floodplain Delineation from Digital Terrain Models.' *Journal of Water Resources Planning and Management,* 129(5), pp. 427-436.

Quan,L. J., and J. Sharma (2005). 'Oracle Application Server 10*g* (10.1.2) MapViewer.' [On-line] November 10 2006. http://www.oracle.com/technology/products/mapviewer/pdf/mapviewer10g_technical_white_paper.pdf

Schwarzenegger, A., and R. R. Renteria (1997). 'Legal Guidelines for Flood Evacuation.' Governor's Office of Emergency Services. California. [On-line] June 08 2007. http://www.oes.ca.gov/Operational/OESHome.nsf/PDF/Legal%20Guidelines%20for%20Flood%20Evacuation/$file/Feat2.pdf

Simonovic, S. P., and S. Ahmad (2005). 'Computer-based Model for Flood Evacuation Emergency Planning.' *Natural Hazards,* 34(1), pp. 25-51.

Zipf, A., and R. Leiner (2004). 'Mobile Internet GIS based Flood Warning and Information Systems.' *2nd Symposium on Location Based Services and Tele-Cartography, Vienna, Austria (January 2004).*

14. Multi-user tangible interfaces for effective decision-making in disaster management

Harmen Hofstra, Henk Scholten, Sisi Zlatanova and Alessandra Scotta

Abstract

In order to handle disasters, save human lives, and reduce damages, it is essential to have quick response times, good collaboration and coordination among the parties involved, as well as advanced techniques, resources, and infrastructure. The current response to disaster is inefficient and sometimes poorly organized. Various studies, research, and analyses have helped clarify needs, understand failures, and improve technology and organizational procedures. One of the major bottlenecks mentioned in many reports is communication between different parties involved in managing a disaster. The lack of a good overview of the locations of teams, personnel, and facilities or insufficient information about the tasks needing to be completed may lead to misunderstanding and errors. Many command and control systems have been developed to aid the decision-making process, but these systems generally require well-trained personnel. This chapter discusses a series of usability investigations completed on a new type of hardware called a 'tangible table' for sharing information and decision-making during emergency responses. The tangible table offers a simple user interface that can be manipulated with human fingers. The interface is intuitive and easy to understand and use. This chapter presents the technology and discusses the interface and the usability tests that have been carried out with different groups of users. The results of these tests show convincingly that the system is highly appropriate for a broad group of non-technical users.

14.1 Introduction

The threat of natural and man-made disasters like earthquakes, plane crashes, or terrorist attacks is ever-present in our society. Past investigations have clearly shown that the number of natural disasters, as well as the victims and economic losses they cause, is rising (http://www.orchestra.org). This requires improvements in disaster preparedness, disaster mitigation, and organization of disaster responses. The emergency response process remains inefficient and sometimes weakly organized (Kevany 2005, Kevany 2008, Winter et al. 2005, Zlatanova et al. 2005, Zlatanova et al. 2007). Over the past few years, many studies, investigations, and analyses have been performed to identify problem areas and suggest solutions. For example, studies in the Netherlands have revealed that the major technology challenges in a response involve collaboration and sharing of information (van Borkulo et al. 2005, Grothe et al. 2008):

- Insufficient collaboration between the parties involved;
- Limited real-time information about the disaster site; and
- Insufficient standardization of data, protocols and services for sharing information

Currently, there is no system of hardware and software that can provide the most appropriate data to the people who need it. Technical support of the decision-making process is especially critical. As will be discussed later, the decision-making process may involve a broad range of users with limited technical experience, but most systems require experts to manipulate them.

Various possibilities can be investigated to enhance and improve the collaboration and decision-making process. In this chapter, we present and evaluate a new type of hardware called the Multi-User Tangible Tabletop Interface (MUTI). MUTI involves much more than touch-sensitive screens or whiteboards shaped like tables. For example, touch screens recognize only single touches and therefore multiple users cannot interact with the system at the same time.

Several different MUTI systems are currently available on the market. Two of them seem to be most promising for emergency response purposes, as Section 3 will discuss in more detail: the Diamond Touch Table (DTT), developed by Mitsubishi Electric Research Laboratories (http://www.merl.com/projects/DiamondTouch/); and the Surface, developed by Microsoft (http://www.microsoft.com/surface/index.html).

The DTT has already been tested in numerous studies examining digital table systems. The DTT is a front-projecting table. This means that the image is displayed with a projector hanging above the table (Fig. 1). The touch signal is recognized with an array of antennas embedded in the touch surface. Each antenna transmits a unique signal, which is received by only one user receiver. Each receiver is connected to the user capacitively, through the user's chair. When a user touches the surface, antennas near the touch point transmit a signal through the user's body into the receiver. This technology allows for multiple touches by a single user and distinguishes between simultaneous inputs from multiple users.

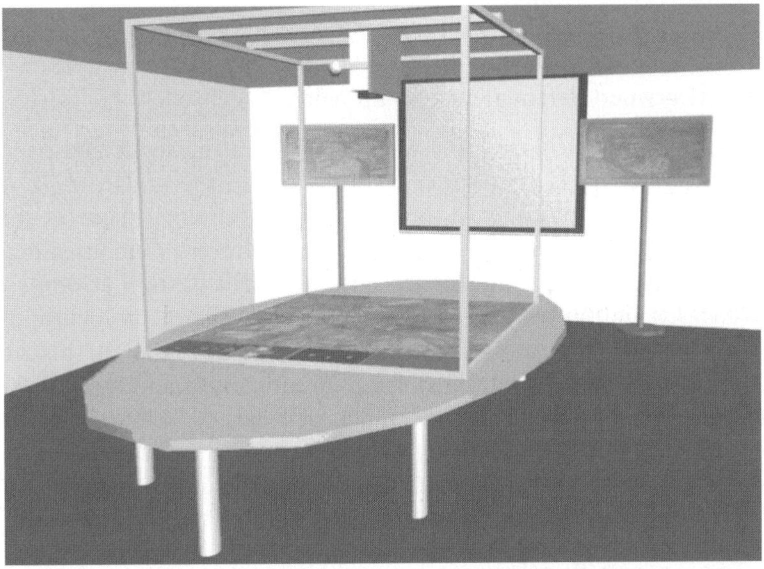

Fig. 1. General set-up of a top-table multi-user tangible interface

Microsoft recently introduced the Surface. The Surface is based on the principle of *frustrated total internal reflection* (http://www.billbuxton.co,/multitouchoverview.html). Infrared LEDs are placed along the edges of a transparent acrylic surface. The infrared light is beamed inside the acrylic and reflects internally. As soon as a finger touches the surface, the internal reflection of the infrared (IR) light is interrupted. Using an IR-sensitive camera, the light bulbs are translated into positions on the screen. In contrast to the DTT, this system does not 'remember' what is drawn by whom.

This next section of chapter concentrates on the functionality that MUTI offers and evaluates it against the tasks needed for an emergency disaster management centre. Tests to examine the ability of DTT to aid the decision-making process in an emergency command and control centre in the Netherlands are discussed in detail.

14.2 Task technology fit for collaborative work

Analysis of collaborative activities shows that people collaborating over a tabletop use specific activities to communicate and structure their activities (Tang 1991). An analysis of these collaborative activities provides guidelines for developing tabletop technology. These guidelines may be (Scott et al. 2003):

- natural interpersonal interaction,
- transition between activities, or
- transition between personal and group work.

It is also important to know for which collaborating tasks and users the tabletop configuration is best suited. This is called the *task-technology fit* of a system. Task-technology fit theories are contingency theories that argue that the use of a technology may result in different outcomes depending upon its configuration and the task for which it is used (Dennis et al. 2001). Research on the subject of task-technology fit for co-located group systems is limited. Nevertheless, several theories do exist for distributed systems. In this chapter, the definitions of Zigurs and Buckland (1998) will be discussed with respect to the MUTI. In general, the task-technology fit consists of three steps: *classifying collaborative tasks*, *classifying MUTI technologies,* and *fitting the MUTI technology* to the defined tasks.

14.2.1 Task classification

Tasks can be classified in different ways. Campbell (1988) proposed a framework based on the importance of task complexity. Any objective task characteristic that implies an increase in information load, information diversity, or rate of information change can be considered to contribute to this complexity. Campbell defines four basic characteristics that meet these requirements. With the presence of each of these characteristics, the level of complexity increases:

- The presence of *multiple potential paths to arrive at a desired end state* increases the information load and thus the complexity. This is true only when one or few of all possible paths lead to the attained goal.
- *Multiple Outcomes*: When the number of desired outcomes increases, the complexity increases. Each possible outcome requires attention. For each possible outcome, a different so-called ‘information processing stream’ is required. Thus, complexity increases with the number of possible outcomes.
- *Conflicting interdependence among paths* occurs when achieving one desired outcome conflicts with achieving another desired outcome. This increases complexity. Campbell (1988) illustrates this type of task with the choice between quality and quantity.
- *Uncertain or Probabilistic Linkages*: Information processing (and thus complexity) increases when it is uncertain whether the chosen solution will lead to the desired and expected outcome.

These characteristics can be used to classify the different tasks into five groups: *simple*, *problem*, *decision*, *judgment,* and *fuzzy* tasks. Simple tasks contain none of the complexity characteristics identified above. Simple tasks have a sole desired outcome, a sole solution scheme, and no conflicting interdependence or outcome uncertainty. Problem tasks involve a multiplicity of paths leading to a well-specified, desired outcome. These tasks are labelled problem tasks because they involve finding the best way to achieve the outcome. Decision tasks emphasize choosing or discovering an outcome that optimally achieves multiple desired end-states. Judgment tasks rely on information sources. The judgment task involves (1) determining which pieces of information are important, (2) weighing these pieces against one another appropriately, and (3) combining the weighted information to arrive at an overall judgment. Fuzzy tasks have both multiple desired end-states and multiple ways of attaining each of the desired outcomes.

14.2.2 Technology classification

Zigurs and Buckland (1998) define Group Support Systems technology as 'a set of communication, structuring, and information processing tools that are designed to work together to support the accomplishment of group tasks.' Most classification schemes for group support systems describe the technology in terms of dimensions, i.e. *communication support*, *process structuring support,* and *information processing support*. Communication support is any aspect of the technology that supports, enhances, or defines the capability of group members to communicate with each other. Process structuring is any aspect of the technology that supports, enhances, or defines the process by which groups interact. Information processing is the capability to gather, share, aggregate, structure, or evaluate information, including specialized templates such as stakeholder analysis of multi-attribute utility analysis.

Technologies can be classified based on how much they support each of these three dimensions. Nevertheless, this framework does not provide a clear method for defining what tools are necessary to support the three dimensions. Recent research on group task usability represents group tasks as low-level actions that a group must carry out in order to accomplish a task in a collaborative fashion (Gutwin and Greenberg 2002). These low-level actions are called the *mechanics of collaboration* and cover two general types of activities: communication and coordination. The mechanics of collaboration are the small-scale actions and inter-actions that group members must carry out in order to accomplish a task collaboratively.

These two categories have been added to the framework presented by Zigurs and Buckland (1998). Since the mechanics of collaboration are independent of the technology and tasks, the framework will be suitable for distributed group support systems as well as for co-located technologies like MUTI technology. The third dimension, information processing, does not apply to the low-level actions in the framework of Gutwin and Greenberg (2000); rather, it depends specifically on the domain in which the system will be used. Tasks that rely on information processing must be supported with software. Examples of information processing in the domain of disaster management include calculating the time to arrive at a certain place, showing information for inhabitants in an affected area, and showing the locations of field actors.

14.2.3 Mechanics of collaboration

The mechanics of collaboration can be further described by differentiating between explicit communication and consequential communication (Baker et al. 2001). Explicit (intentional) communication is that which is planned, such as 'conversation' and 'gestures,' or 'written communication,' 'deictic references,' and 'manifesting actions' (Pinelle et al. 2003). Verbal communication (conversation) serves the communication process in two ways. First, people explicitly talk to each other to tell them what they are doing. Second, people can pick up commentary that is intentionally produced alongside their actions, called verbal shadowing (Gutwin and Greenberg 2002). Gestures are frequently used in communication and are important in communicating messages to others. Examples include pointing at specific objects or using hand gestures to clarify a message. Written communication can be used to communicate with someone who is not present at that moment or to annotate or provide details about items (Pinelle et al. 2003). Deictic references are combinations of speech (verbal communication) and gesture communication. In contrast, manifesting actions replace verbal communication entirely: picking up the phone indicates that someone is going to make a call.

The consequential communication is information that comes from a person who does not intentionally act to inform the other person. The information source is the other person's bodily actions in the workspace. Most things that people do in a workspace are done through some kind of bodily action, for instance changing body position, moving the heads or arms, or looking in a certain direction (Zigurs and Buckland 1998}. Watching other people work is an important source of information in communication. This differs from intentional gestures in that in consequential conversation the producer of the information does not intentionally act to inform the other person. Another form of consequential communication involves gathering information on who is collaborating, which can be done by observing who is in the workspace, what they are doing, and where they are working. A third form of consequential communication is the use of artefacts and feedback. Tools have particular sounds, like the snip of scissors or the scratch of a pencil, but the feedback of an artefact provided by an interface can also provide valuable information to users and people in the area. For instance, a sound is used to indicate 'affirmative' when a command is given in WarCraft-III (Gutwin Greenberg 2002).

Process structuring is defined as any aspect of the technology that supports, enhances, or defines the process by which groups interact (Zigurs and Buckland 1998). Process structuring activities involve planning work and coordinating activities. The mechanics derived from planning work generally have three parts: obtaining a resource, reserving a resource for future use, and protecting work. Transferring objects and resources are considered coordinating activities (Pinelle et al. 2003). This activity can involve two type of mechanical action: handoff of objects and deposits.

Obtaining a resource is important in collaboration because objects or areas in the workspace are often only available for one person at a time. An example of reserving objects for future use is reserving areas of the workspace for later use. Protecting work in this way can prevent others from destroying or altering work done by an actor. Physically or verbally giving or taking objects (handoff objects) and placing objects for notification (deposit) are critical for moving objects and tools between people (transfer).

14.3 Multi-user tangible interface

This section focuses on MUTI technology and defines the level of support for each of the three dimensions as defined in the previous section. First, the defined mechanics will be broken down into typical activities. MUTI technology will then be analyzed to determine the level of support for the identified activities.

14.3.1 Activities

Two categories of activities within the dimension of communication support have been identified: explicit communication and consequential communication. A set of typical actions needed to support these activities has been derived. For intentional communication, the activities are *explicit talk*, *written messages*, *intended gestures* (subdivided into indicating, drawing, demonstrating), and *deictic references* (pointing and conversing). For consequential communication, the activities are *overhearing others' conversations*, *observation* (of who is around, what people are doing, where people are working), *use of artefacts,* and *feed through* (subdivided into notification of changes to objects/items), *observation* (of body position and location, and direction of gaze), and *feed through* (notify changes to objects). The level of support for these actions depends on the physical configuration of the system and requires information about other users around the table and the workspace. Therefore, the process of information gathering, or consequential communication, also needs to be supported.

In general, the dimension of process structuring consists of five types of activities. These activities involve a set of typical actions: *reserving areas* of the workspace, *protecting work from alterations*, *observation* (of who is around, what they are doing, where they are working), *preventing conflicts* between different people's work, and *reserving objects* for future use.

Many of the identified activities based on the mechanics of collaboration depend on awareness. In the field of Computer Supported Cooperative Work, researchers have proposed four types of awareness that apply more specifically to groups working face to face (Gutwin et al. 1996). *Informal awareness* is knowledge about who is around and what he or she is doing. *Social awareness* is keeping track of communicational information about others, such as whether someone is paying attention or their level of interest. *Group structural awareness* is information about people's roles, responsibilities, or status. The fourth type of awareness is *workspace awareness* and is defined as perspective of one worker observing/interacting with others.

No typical collaborative activities apply to the dimension of information processing. This dimension relies on the software architecture and software features of the system. The level of support for each of the dimensions depends on the physical configuration, the technical configuration, and the system capabilities.

14.3.2 The technology

The physical configuration of a MUTI is based on a tabletop (Fig. 2). This means that people are positioned face to face around a shared display. The main difference from distributed group systems is that people not only share a virtual workspace but they also share the same physical location and are thus able to see each other.

Fig. 2. Diamond Tangible Table, tested at Geodan, the Netherlands

The face-to-face configuration of MUTI technology naturally supports *informal awareness*, *social awareness,* and *group structural awareness*. As mentioned above, many of the actions described depend on these types of awareness. Some of the actions, however, rely more on workspace awareness or require additional (software) support.

Another important aspect of the technology is the *size of the table*. Although the size of the table is known to influence the way people collaborate, research on this question is limited. Ryall et al. (2004) were unable to identify significant effects of changing table size on the speed of task completion. However the authors did observe a link between the size of the table and social interaction, physical reach, and visibility of the workspace. People must be close enough to communicate but must have their own private space. Hall (1966) reports that people generally feel comfortable working at an arm's length since this preserves their personal space. Workspace awareness, however, requires people to be close enough to the workspace so that the actions inside the workspace are visible to all users. The mechanics of assistance require that all users be able to reach the whole workspace. Thus, the size of the table should allow all users to reach the whole workspace and see all actions of other users. This increases workspace awareness but may cause people to work within the private space of other actors. Therefore, Scott et al. (2003) state that in order to support different kinds of tabletop activities, the technology must be flexible enough to allow users to interact from a variety of positions around the table.

One important issue in communication and carrying out collaborative tasks is the choice of *input device*. Two types of input device are *indirect input devices* such as a mouse, and *direct input devices* such as touch and speech. A study on direct and indirect input devices and their effects on collaboration showed that direct input on tabletop displays supports natural gesturing and allows users to easily notice their partner's actions. In addition, direct tabletop input can allow rich interpersonal interactions, enabling users to both convey and understand one another's intentions seamlessly (Whalen et al. 2004). Although a user may be more familiar with indirect input devices, direct input devices such as touch support the mechanics of collaboration and make it easier to take full advantage of the technology. Keeping track of many mice is nearly impossible. This leaves users physically pointing at their virtual pointers to tell other users where they are. In addition, using separate physical devices keeps users from engaging in the natural human tendencies to reach, touch, and grasp.

Fig. 3. Orientation of objects towards the users

Another critical feature of MUTI is *feedback* or *process feed through*. Tools should have particular sounds, such as the snip of scissors or the scratch of a pencil, but the feedback of an artefact provided by the tabletop system should also provide valuable information to users and people in the surrounding area.

People located around a shared table workspace do not all see things with the same *orientation* (Tang 1991). Kruger et al. (2003) identified three key roles of orientation that affect collaboration and therefore affect the design of tabletop interfaces: *comprehension*, *coordination*, and *communication*. Comprehension is the orientation of an object in order to interpret something such as text or symbols. This form of orientation is used to position an item in such a way that it is easy to read, or to provide the best angle for completing a given task (Fig. 3).

Another feature of the MUTI is that the surface should be *debris tolerant*. This means that the table will not react when people put things on it or even spill liquids on it, as on a normal table.

14.3.3 Level of support

In this section, we analyze activities and the underlying actions that can be supported by the MUTI technology. A distinction will be drawn between tasks supported by the physical configuration of the technology and tasks that can be supported by software development. The level of support is determined as low, medium, and high. The first group of activities is related to the dimension of communication. Table 1 illustrates the identified level of support and the functionality offered by the software.

Table 1: Level of support for different communication processes

	Hardware support	Software support
Explicit talking	High	No
Overhearing others	High	No
Writing messages	Low/medium	Yes
Indicating	High	Yes
Feedback giving	High	Yes
Deictic referencing	High	Yes

Explicit talk and verbal shadowing require that users be able to convey a message to a listener. Conversational speech also requires some indication that the message has been received and understood (Pinelle et al. 2003). Explicit talk and verbal shadowing rely on social and informal awareness. The face-to-face configuration of MUTI technology naturally supports these forms of awareness. Written messages on a shared tabletop interface must be oriented properly for users to be able to read them. The fact that the table allows objects to be placed on the table without interfering with the activities does make it possible to write messages on sheets of paper. Indicating/pointing, demonstrating, and drawing are gesture activities and require that people be able to see one another's gestures. The face-to-face configuration of the Diamond Touch (Fig. 3) naturally supports this activity. The level of support with Diamond Touch for this activity is therefore high.

The activity of noticing who is around, what people are doing and where people are working constitutes information awareness, which arises naturally from the face-to-face configuration of the MUTI technology. This information-gathering activity is also supported through the use of touch as a direct input device. People orient work materials toward themselves, indicating to others that they are working with those particular materials (Fig. 4).

Fig. 4. Two people working on different objects (courtesy of Wu et al. 2006)

Deictic activity is the act of verbal communication and the use of artefacts to support this communication visually. The artefacts act as conversational props. This is supported by the face-to-face configuration and the physically shared single workspace.

In contrast to activities in the communication dimension, the level of support for is not high for all activities in the dimension of process structuring. The fact that people collaborate over a single display makes it more difficult to prevent conflicts with other users working on the single display. Table 2 illustrates these findings:

Table 2: Level of support for different activities in the dimension of process structuring

	Hardware support	Software support
Reserving areas	Medium	Yes
Protection	Medium	Yes
Observation	High	Yes
Conflict prevention	Medium	Yes

Space can be reserved explicitly, for instance by informing other users through speech or through implicit communication. Although the MUTI technology does support this action, reserving space does require significantly more attention compared to traditional desktop applications. Therefore, the level of support for this action is defined as medium.

Protecting objects from alteration by other users requires the MUTI technology to identify the different users of the system. The DTT identifies which user is touching where. As mentioned above, this is accomplished by signals transmitted to antennas coupled to specific users. Another form of protection is through ownership of objects, by orienting and placing objects in a way that it is clear to other users who 'owns' the object (Fig. 4). However, the fact that users are working on the same interface hampers this, so the level of support for this mechanic is defined as medium.

The activity of observing receives a high level of support, because of the format of face-to-face work. At the same time, working on a single display makes it difficult to prevent conflicts with other work. However, reserving areas of the workspace and protecting them from alteration can prevent these conflicts. Therefore the level of support for this action is defined as medium.

We have used the framework of Zigurs and Buckland (1988) to determine the level of support for specific tasks according to the level of support for each of the three dimensions discussed above (Table 3). The overall conclusion is that MUTI technology provides added value for tasks involving highly complex tasks, especially fuzzy tasks.

Table 3: Level of support for each dimension as defined by task type

Task Type	Communication	Process structuring	Information processing
Simple	High	Low	Low
Problem	Low	Low	High
Decision	Low	High	High
Judgment	High	Low	High
Fuzzy	High	Medium	High

14.4 Applicability to disaster management tasks

Response to a disaster consists of many different actors who carry out many different activities in many different settings. Not all settings can benefit from MUTI technology. For instance, firemen extinguishing a fire on the field will definitely not benefit from the technology. The level where the technology is most likely to add value is the tactical level (Borkulo et al. 2005, Diehl and Heide 2005, Hofstra 2006), where coordinating and implementing the decision-making process, as well as harmonizing and coordinating between actors takes place.

In the Netherlands, the Regional Operational Team (ROT) is the unit that operates at the tactical level. Each emergency response sector involved in a particular emergency response has one representative in the ROT: there is representative each from the fire brigade, the medical service, and the police. The Operational Leader (OL) is in charge of the ROT. Within the ROT, strategic and tactical decisions are translated into operational assignments for the sectors involved in the disaster response.

Investigating the applicability of the MUTI for disaster management requires classifying the types of task to be performed within the ROT. The main task on a tactical level is planning the crisis response process. Gervasio and Iba (1997) define three main themes in times of crisis: *threat*, *urgency,* and *uncertainty*. These three themes will be further examined to define the characteristics of tasks involved in disaster management planning.

In the response to a disaster, threat is ever-present. When a disaster occurs, there is a threat of losing things of great value, namely people or assets. Not having a clear goal makes the tasks far more complex. For instance, the disaster manager should be able to make a choice between immediate evacuation or fighting the source of the disaster in order to ensure the safety of people and assets. Another example of high complexity is when multiple possible actions lead to conflicting outcomes. For example, evacuating people conflicts with protecting people at the site of the incident. Urgency refers to the limited time available to make decisions. This limited time makes it difficult and sometimes impossible to consider all possible options. Moreover, in the first few hours after disaster strikes, information is very limited. This leads to uncertainty about what is happening and therefore to uncertainty about outcomes of actions.

Consequently, the conclusion is that these tasks can be considered highly complex and therefore 'fuzzy tasks.' Several studies classify and discuss further tasks and processes in emergency response in the Netherlands (Borkulo et al. 2005, Snoeren et al. 2007, Diehl et al. 2006). This chapter will focus on an empirical study looking only at the use of DTT to support the work of the ROT.

14.5 Empirical study

An empirical study would ideally be carried out with professionals working in the disaster management field, but this is usually very difficult. Therefore other approaches are necessary to field-test MUTI and related technologies.

The actions that need to be carried out in response to a disaster depend strongly on the type and impact of the disaster. In the Netherlands, 25 different processes are strictly defined, each as part of a cluster. These clusters are classified according to the sector that is responsible for the process (e.g., fire brigade, police, medical service, and police). From all the processes, *Observations and Measurements* was selected to test MUTI applicability. This process is the responsibility of the fire brigade and comprises all the activities for performing measurements in an area affected by a release of dangerous substances. This process was chosen because it not only contains activities on the tactical level of crisis response but it also includes planning activities. The first step in understanding this process is to break down the crisis response process into the activities of the different actors. Notation of the activities has been carried out using the unified modelling language (UML) by applying use-case and activity diagrams. These diagrams provide a clear view of the users involved and the tasks they must perform. The UML diagrams are explained in detail in Hofstra (2006).

To evaluate the user acceptance of MUTI, a Technology Acceptance model (TAM) was applied (Davis 1989). TAM was built based on the theory of reasoned action (TRA). The TRA is based on the hypothesis that if a person intends to behave in a certain way, it is likely that the person will act as intended. According to the TRA, a person's intention is determined by two things: first, the attitude towards the behaviour, and second, the subjective norm. This subjective norm is the way a person thinks other people would view him or her if he or she performed a certain behaviour. The TRA is an intentional model that has proven successful in predicting and explaining behaviour across a wide variety of situations.

Venkatesh et al. (2000) reported that TAM gives the best results for testing user acceptance of (new) technologies. The TAM model focuses on fully functional systems implemented and tested in existing organizations involving real (potential) users of the system. In our case, we investigated

MUTI technology prior to developing a fully functional system, so the results presented here evaluate only some features of the MUTI technology

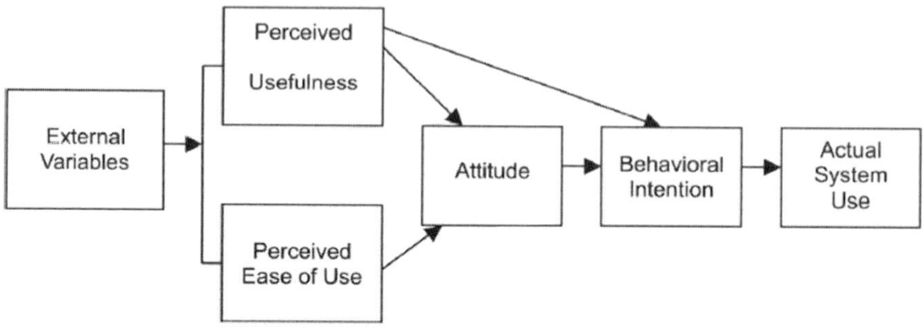

Fig. 5. Technology acceptance model (Davis 1989)

Figure 5 shows a schematic view of TAM. The arrows represent the relationships among the major components of the model. The attitude towards using the system determines the behavioural intention to use and behavioural intention to use determines actual system use. These relationships are based on the TRA, according to which a person's intentions are defined by his attitude towards the behaviour and the behaviour is based on a person's intentions to behave in a certain way. If a person positively values an outcome, this feeling can often increase one's commitment to behave in a way that achieves that outcome. Ease of use is also hypothesized to have a significant effect on attitude towards use. The easier a system is to interact with, the greater the user's sense of efficacy and personal control.

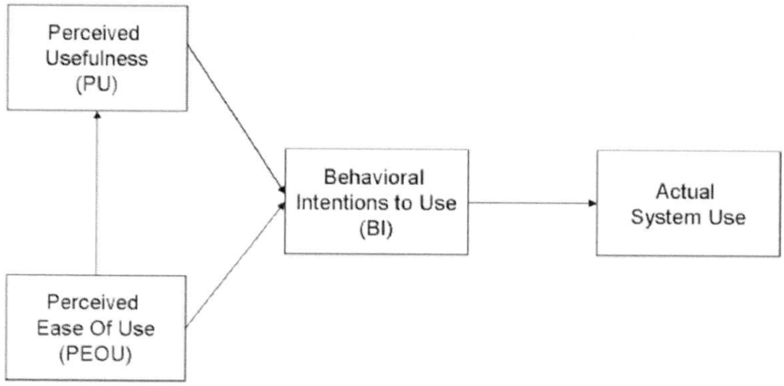

Fig. 6. Simplified model used in this study

For simplicity, our study uses a simplified TAM without the component of "attitude toward behaviour" (Fig. 6). This simplification has already been shown to give good results in other studies, e.g. Venkatesh et al. (2000).

14.5.1 Test set-up

We applied this simplified theory in our study by preparing a questionnaire and a scenario.

The most common way to collect data for TAM studies is by administering questionnaires. Therefore the questions must be designed in such a way that they translate into relationships. On the basis of previous studies, a scale for usefulness and perceived ease of use were developed. The participants responded to questions about themselves on a paper questionnaire (self-administration). This type of survey was chosen mainly because of its rapid administration but also because it ensured anonymity and privacy for the participants and thereby increased the likelihood of getting honest responses. Perceived Usefulness (PU) and Perceived Ease Of Use (PEOU) were measured using the seven-point Likert scale from (1) 'strongly disagree' to (7) 'strongly agree.' Two questions in the survey referred to the 'intention to use' constructs. These questions allowed testing of the effect of the PU and EOU constructs on the actual intention to use. This provided an indication of how well the theory of reasoned action, and therefore TAM, applied to this study. Responses to these two questions were also measured on a seven-point Likert scale.

As mentioned above, one of the Dutch emergency response processes was used for the scenario, as the emphasis was on geo-information and GIS tools. The software used for the experiment was designed and developed by Geodan (www.geodan.nl) in cooperation with major parties involved in a project on disaster management in the Netherlands (www.gdi4dm.nl). The two basic features tested were zooming and panning of a map.

The users participating in the experiment were employees of Geodan and emergency responders. Geodan is a company that develops geographical information systems for a variety of purposes. Most of the participants in the experiment could be considered professionals in the field of GIS. The test session took place in an artificial, controlled environment at Geodan. After a short explanation of the functionalities (zooming and panning), the participants were asked to carry out a prepared assignment on a map of Amsterdam. After the assignment was completed, the participants filled out the questionnaire.

14.5.2 Results

Thirty-five people participated in the test. The 35 respondents consisted of 29 males and 6 females. Most were between 25 and 40 years old; four respondents were older and one was younger. The level of education of the respondents was High Technical or University, except for one. One subject reported a 5 on the 6-point scale regarding previous experience with the Diamond Touch Table or similar technologies; this subject did not meet the inclusion criteria of the user group and was therefore removed from the study. Thus, the total number of participants was 34.

The variables of education and job description were rescaled to create comparable groups (Fig. 7). The participants were subdivided into four different groups according to their background: non-technical background, non-geography background, performing non-technical tasks, and performing non-geography tasks.

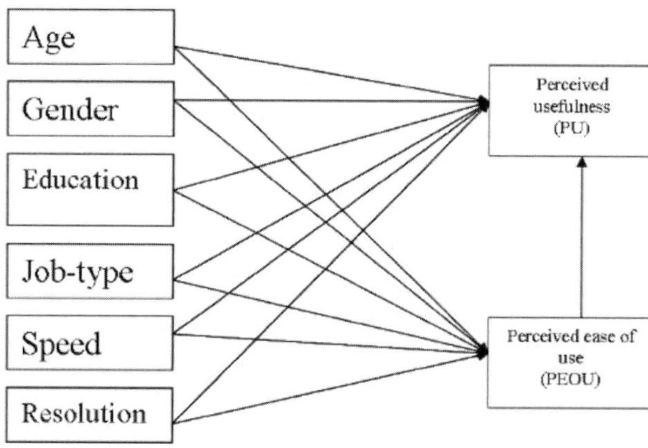

Fig. 7. External variables.

The results of the test were evaluated with respect to *reliability* and *inner construct correlation.*

Reliability is defined as 'the correlation between answers to questions that measure the same construct.' Cronbach's Alpha is a common measure of reliability of a psychometric instrument (Fig. 8) and is therefore useful in TAM studies.

$$\alpha = \frac{N \cdot \bar{r}}{1 + (N-1) \cdot \bar{r}}$$

Fig. 8. Cronbach's Alpha: *N* is the number of components and *r* is the average of all (Pearson) correlation coefficients between pairs of components.

Cronbach's Alpha increases when the correlations between the items increase. Table 4 shows the alpha for the constructs measured in this test. The higher the Alpha is, the more reliable the test; a value of 0.7 or greater is acceptable (Nunnally et al. 1978). The alpha of all constructs in this test was very high. This is not surprising, since the questions used here have been validated in several previous studies (Table 4).

Table 4 : Reliability analysis

Construct	Cronbach's Alpha	N of items
Usefulness	0.89	10
Ease of use	0.85	4
Intention to use	0.92	2
PU panning	0.92	5
PU zooming	0.80	5

The inner construct correlation reflects the degree to which the variables are related. The most common measure is Pearson`s correlation. Pearson`s correlation reflects the linear relationship between two variables. The statistic is defined as the sum of the products of the standard scores of the two measures, divided by the degrees of freedom. It ranges from +1 to -1; zero indicates that no relation is discovered. The correlation between the measured constructs is displayed in Table 5 and Fig. 9. All relations between the measured constructs were found to be positive. This correlation was expected since the TAM model has been used and validated in many previous studies. The positive correlations in this empirical study show that the perceived usefulness and perceived ease of use are useful for predicting the intention to use a MULTI.

Table 5: Pearson Correlations between PU, PEOU and intention to use (ITU)

		PU	ITU	PEOU
Perceived Usefulness	Correlation	1	0.524	0.699
	Significance		0.001	0
Intention to use	Correlation	0.524	1	0.524
	Significance	0.001		0.001
Perceived ease of use	Correlation	0.699	0.524	1
	Significance	0	0.001	

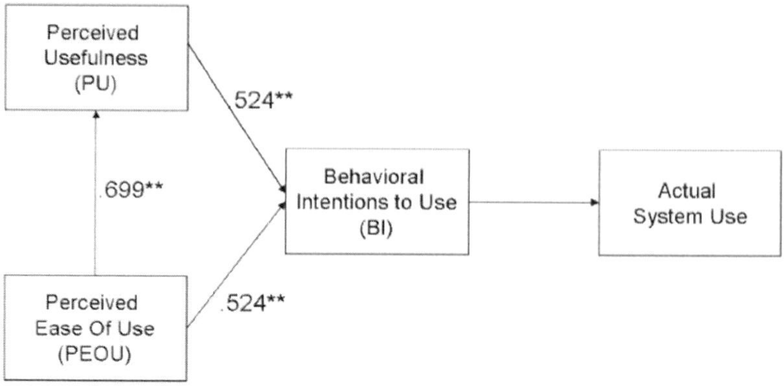

Fig. 9. Inter-construct correlations

The results of perceived usefulness and perceived ease of use were rescaled to range from zero to six. Zero corresponds to 'completely disagree' and six to 'completely agree' (0 = strongly disagree, 1 = disagree, 2 = somewhat disagree, 3 = undecided, 4 = agree, 5 = somewhat agree, 6 = strongly agree). The rescaling is given in Table 6.

Table 6: Perceived usefulness and ease of use

	N	Mean	Std deviation	Variance
PU	34	4.4882	0.8789	0.773
PUOU	34	4.4412	1.002	1.004

The results show no relationship between the age of the participants and the usefulness or the ease of use of the MUTI. This may be explained by the fact that most respondents were between 25 and 40 years old. Since only six participants of 34were female, no conclusion could be drawn about the influence of gender.

No significant relationships were found between participant background and the perceived usefulness or ease of use of the system. Whether someone has geographical education or not does not seem to make any difference in how useful or easy to use they find the system. Similarly, whether or not a participant has followed a technical course of study does not influence these variables. Finally, we found no relationship between the type of job and the reported usefulness or ease of use. We note that job type was tested for only two of the four groups: technical or non-technical jobs, not for GIS or non-GIS jobs. The results clearly show that no technical knowledge is required to work with the MUTI. Whether knowledge of geographical information systems influences the usefulness of the tool and the ease of use could not be concluded

The speed of the system was measured for zooming and panning of a map. Users who rated these features as faster than working with a mouse also rated the feature as more useful. Interestingly, users who rated the features as faster did not find the system to be easier to use. Thus, although the software used for the experiment was somewhat slow in rendering new maps, this did not negatively affect participants' perception of the tool's ease of use.

We observed similar results for the quality of the maps. Maps with a higher resolution made the system more useful but not easier to use. More experienced GIS users rated the higher resolution maps lower compared to less experienced GIS users. This can be explained by the fact that more experienced GIS users have higher expectations regarding the quality of the maps. More details on external variables can be found in Hofstra (2006).

14.6 Conclusions

In this chapter, we have discussed our study of the usefulness of the Multi-User Tangible Tabletop Interface for disaster management. Our theoretical study and empirical testing with a group of 35 participants clearly show the applicability of this technology to some of the processes in disaster management in the Netherlands. Test participants responded positively in their evaluation of the usefulness and ease of use of the Diamond Tangible Table and the software developed to carry out basic GIS tasks.

Disaster management--and emergency response more specifically--is a very specific type of application that involves many users with different backgrounds and various responsibilities. These users must interact with each other and be able to discuss possible solutions to determine the best solutions. The new technology has revealed various promising features that may provide solutions to many drawbacks of existing desktop systems (screen, keyboard, and mouse). Tangible technology offers better options for group work, specifically for decision-making at the tactical level in command centres.

Many tasks in disaster management require a geographical component; for example, using maps is critically important for situational awareness of rescue units, victims, and civilians in danger. In this respect, the natural hand-based MUTI allows users to point and discuss, as with a paper map, but the system has all the advantages of digital screens (e.g. zoom in, zoom out, pan). When used for basic GIS tasks, this interface is very likely to be accepted by users.

The system was tested using the simplest interactions that are performed every day with a mouse, i.e. click, drag, and mouse-over. Thus the testing allowed for the completion of a few operations, such as object selection, zooming in/out and panning of maps, and simple object drawing. More complex activities, however, such as gesture recognition, were not tested.

The potential of this technology for practical application is very high. Theoretically its visualisation hardware can be integrated into any system architecture. In this respect, an important next step will be adapting and extending available command and control system (CCS) software and its dynamic database for use on the DTT. Further developments such as these promise to meet the needs of Regional Operational Teams, e.g. monitoring vehicles and people, analyzing and choosing suggested routes, monitoring plumes, and performing spatial analysis using tools such as buffer and overlap.

References

Baker, K., S. Greenberg, and C. Gutwin, 2001, Heuristic Evaluation of Groupware Based on the Mechanics of Collaboration. Proceedings of the 8th IFIP International Conference on Engineering for Human-Computer Interaction, pp. 123-140

Berlo, M. van, R van Rijk and E. Buiël, 2005, A PC-based Virtual Environment for training team desicion-making in high-risk situations, in: Proceedings of the 2nd International ISCRAM Conference April 2005, Brussels, Belgium, pp. 195-201

Borkulo, E van, H.J. Scholten, S. Zlatanova and A van den Brink, Decision making in response and relief phases, in: PJM van Oosterom, S Zlatanova & EM Fendel (Eds.), Geo-information for disaster management - late papers, pp. 47-54.

Borkulo, E, V. S. Barbosa, A. Dilo, S. Zlatanova and H. Scholten. 2006, Services for an Emergency response system in the Netherlands, in 'Proceedings of the Second Symposium on Gi4DM', 25-26 September, Goa, India, 6 p.

Carter, E. and S. French, 2005, Nuclear Emergency management in Europe: a review of approaches to decision making, in: Proceedings of the 2nd International ISCRAM Conference April 2005, Brussels, Belgium, pp. 247-259

Campbell, D., 1988, Task complexity: a review and analysis, tThe Academy of Management Review, 13(1):40-52.

Davis, F., 1989, Perceived Usefulness, Perceived Ease of Use, and User Acceptance of Information Technology. MIS Quarterly, 13(3):319-340

Dennis, A., B. Wixom., and R. Vandenberg, R, 2001, Understanding fit and appropriation effects in group support systems via meta-analysis. MIS Quarterly, 25(2):167-193

Diehl, S. and J. van der Heide, 2005, Geo Information Breaks through sector shink, in: PJM van Oosterom, S Zlatanova & EM Fendel (Eds.), Geo-information for disaster management, Springer Verlag, Heidelberg, pp. 85-108

Diehl, S, J. Neuvel, S. Zlatanova and H.Scholten, 2006, Investigation of user requirements in the emergency response sector: the Dutch case, Second Symposium on Gi4DM, 25-26 September, Goa, India, CD ROM, 6p.

Fabri, K. and G. Weets, 2005, ICT for Environmental Risk Management in the EU Research Context, in: PJM van Oosterom, S Zlatanova & EM Fendel (Eds.), Geo-information for disaster management, Springer Verlag, Heidelberg pp. 51-56.

Gervasio, M. and W. Iba, 1997, Crisis Response Planning: A Task Analysis. Proceedings of the Nineteenth Annual Conference of the Cognitive Science

Grothe, M.J.M., H.C. Landa, and J.G.M. Steenbruggen, 2008, in: Zlatanova &Li (eds.) Geospatial information technology for emergency response, ISPRS Book series Vol.6, Taylor and Francis, London, pp. 313-328

Gutwin, C. and S. Greenberg, 2000, The Mechanics of Collaboration: Developing Low Cost Usability Evaluation Methods for Shared Workspaces. Proceedings of the 9th IEEE International Workshops on Enabling Technologies: Infrastructure for Collaborative Enterprises, pp. 98-103

Gutwin, C. and S. Greenberg, 2002, A Descriptive Framework ofWorkspace Awareness for Real-Time Groupware. Computer Supported Cooperative Work (CSCW), 11(3):411-446.

Gutwin, C., S. Greenberg, and M. Roseman, 1996, Workspace Awareness in Real-Time Distributed Groupware: Framework, Widgets, and Evaluation. Proceedings of HCI on People and Computers XI, pp. 281-298

Hall, E., 1966, Distances in Man: The Hidden Dimension

Hofstra, H, 2006, Multi user tangible interfaces and disaster management: A study on the added value of multi user tangible interfaces in crisis response, Msc Thesis, VU University Amsterdam, 85 p.

Hoogendoorn, M., C. M. Jonker, V. Popova, A. Sharpanskykh and L. Xu, 2005, Formal modeling and comparing of disaster plans, in: Proceedings of the 2nd International ISCRAM Conference April 2005, Brussels, Belgium, pp. 97-107

Kevany, M. 2005, Geo-information for disaster management: lessons from 9/11, in: PJM van Oosterom, S Zlatanova & EM Fendel (Eds.), Geo-information for disaster management, Springer Verlag, Heidelberg, pp. 443-464

Kevany, M, 2008, Improving geospatial information in disaster management through action on lessons learned from major events, in: Zlatanova &Li (eds.) Geospatial information technology for emergency response, ISPRS Book series Vol.6, Taylor and Francis, London, pp. 3-19

Kruger, R., S. Carpendale, S. Scott, and S. Greenberg, 2003, How people use orientation on tables: comprehension, coordination and communication. Proceedings of the 2003 international ACM SIGGROUP conference on Supporting

Nunnally, J. et al., 1978, Psychometric theory, McGraw-Hill New York. group work, pp. 369-378

Pinelle, D., C. Gutwin, and S. Greenberg, 2003, Task analysis for groupware usability evaluation: Modeling shared workspace tasks with the mechanics of collaboration. ACM Transactions on Computer-Human Interaction (TOCHI), 10(4):281-311

Ryall, K., C. Forlines, C. Shen, and M. Morris, 2004, Exploring the effects of group size and table size on interactions with tabletop shared-display groupware. Proceedings of the 2004 ACM conference on Computer supported cooperative work, pp. 284-293

Scott, S., K. Grant, and R. Mandryk, 2003, System guidelines for co-located, collaborative work on a tabletop display. Proceedings of ECSCW 2003, pp. 159-178.

Scotta, A., H. Scholten, and H. Hofstra, 2008, Multi-user tangible interface for decision making in emergency responses, in: Coors, Rumor, Fenedel&Zlatanova, Urban and regional data management, UDMS Annual 2007, Taylor and Francis, pp. 357-368

Scholten, H.J., A.J. LoCashio & T. Overduin, 1999, Towards a spatial information infrastructure for flood management in The Netherlands, In: Special features in coastal conservation 3, eds. J. Visser & R. Misdorp, Opulus Press, Uppsala, pp 151-160.

Snoeren, G., S. Zlatanova, J. Crompvoets, H. Scholten, 2007,Spatial Data Infrastructure for emergency management: the view of the users, Proceedings of the 3rd International symposium on Gi4DM, 22-25 May, Toronto, Canada, CDROM, 12 p.

Tang, J., 1991, Findings from observational studies of collaborative work. International Journal of Man-Machine Studies, 34(2):143-160

Venkatesh, V., M. Morris, G. Davis, and F. Davis, 2003, User Acceptance of Information Technology: Toward a Unified View. MIS Quarterly, 27(3):425-478.

Whalen, T., V. Ha, K. Inkpen, R. Mandryk, S. Scott, and M. Hancock, 2004, Direct Intentions: The Effects of Input Devices on Collaboration around a Tabletop Display. Dalhousie University.

Winter, S., G. Iglesias, S. Zlatanova and W. Kuhn, 2005, GI for the public: the terrorist attack in London, Geoinformatics Magazine, No. 7, Oct/Nov 2005, Vol. 8., pp18-21

Wu, M. C. Shen1, K. Ryall1, C. Forlines1 and R. Balakrishnan, 2006, Gesture Registration, Relaxation, and Reuse for Multi-Point Direct-Touch Surfaces, IEEE International Workshop on Horizontal Interactive Human-Computer Systems (TableTop), January 2006, pp. 183-190

Zigurs, I. and B.Buckland, 1998, A theory of Task/Technology Fit and Group Support Systems Effectiveness, MIS Quarterly, 22(3):313-334.

Zlatanova, S., A. Fabbri, J. Li, 2005,Geo-information for Disaster Management: Large scale 3D data needed by Urban Areas, GIM International, March 2005, Volume 19, Issue 3, pp. 10-13

Zlatanova, S., D. Holweg and M. Stratakis, 2007, Framework for multi-risk emergency response, in: Tao&Li (Eds.) Advances in Mobile Mapping Technology, Taylor&Francis, London, ISPRS Book Series, pp. 159-171

Index

About the Editors

Dr. Sisi Zlatanova is an associate professor at the GIS technology section, Delft University of Technology, The Netherlands and leading a theme group 'Geo-information for Crisis response'. She has graduated as a surveyor from the University of Architecture, Civil Engineering and Geodesy, Sofia, Bulgaria, in 1983 and obtained her Ph.D. in 3D GIS for urban modelling at the Graz University of Technology, Graz, Austria, in 2000. She has worked as a software programmer at the Central Cadastre in Sofia, Bulgaria, an assistant-professor at UACG, Sofia, Bulgaria, and a researcher at the International Institute for Geo-Information Science and Earth Observation (ITC), Enschede, The Netherlands. Her research interests are related to use of spatial technologies in emergency response. Special attention is given to the third dimension: 3D object reconstruction, 3D data structures and geo-databases, 3D spatial relationships (topology) and 3D visualization (VR and AR). She is author of numerous publications on 3D modelling and technology for emergency response. She has been char of ISPRS IV/8 'Spatial data integration for emergency services' for the period 2004-2008, which has been responsible for the International annual symposium 'Geo-information for Disaster Management' (Gi4DM).

Dr. Shailesh Nayak is Director, Indian National Centre for Ocean Information Services (INCOIS), an autonomous institution under the Ministry of Earth Sciences, Hyderabad, India. He has been providing leadership for realizing and operating on Establishment of National Early Warning System for Tsunami and Storm Surges in the Indian Ocean and conceptualization and development of Marine GIS. He has obtained PhD degree in Geology from the M.S University of Baroda in 1980. He joined the Space applications Centre of the Indian Space Research Organisation in 1978 as a scientist and became Director of Marine and Water Resources Group in 2001. His main research interest was use remote sensing and GIS for coastal and marine ecosystems and management research. He has published more than 70 papers in this field. He has been recipient of the National Mineral Award for the year 2005. He was awarded the Indian National Remote Sensing Award for the year 1994 by the Indian Society of Remote Sensing, Dehradun. He has been the President, International Society of Photogrammetry and Remote Sensing (ISPRS), Technical Commission (TC) IV on 'Geo-databases and Digital Mapping' for the term 2004-08. He was instrumental in organizing the International symposium 'Geo-information for Disaster Management' (Gi4DM) in Goa in 2006.

Printing: Krips bv, Meppel, The Netherlands
Binding: Stürtz, Würzburg, Germany